Utopian Literature

Solaris Farm

Milan C. Edson

ARNO PRESS & THE NEW YORK TIMES
NEW YORK • 1971

Reprint Edition 1971 by Arno Press Inc.

Reprinted from a copy in The Pennsylvania State University Library

LC# 78-154440
ISBN 0-405-03523-3

Utopian Literature
ISBN for complete set: 0-405-03510-1

Manufactured in the United States of America

CAPTAIN MILAN C. EDSON.

SOLARIS FARM;

A Story of the Twentieth Century.

BY

MILAN C. EDSON.

Published by the Author

at

1728 New Jersey Ave., N. W., Washington, D. C.

In the Year 1900.

Press work by Byron S. Adams,

DEDICATION.

This book, is dedicated to the sons and daughters of the farms of the Republic as an expression of the author's realization, that Agricultural people constitute a large majority of its working units: That as such, its destiny is in the hands of their boys and girls, as its future guardians, fathers and mothers: That for the reasons stated, they should become its dominant thinkers and leaders: That Agriculture is the true basis of industrial and commercial success; hence, it should be made the most noble and pleasing of all occupations: That the alarming encroachments of land monopoly, and the inability of the small farm to meet the expense of using the latest and best machinery, threatens the total extinction of all land-owning farmers, and of their consequent reduction to the dependent caste of farm laborers: That the isolated life and the severe toil of the small farm, has a dangerously depressing effect on the minds of its people: That all of these things, seem to demand the changes suggested by the contents of this book.

PREFACE.

Strong in my convictions that all civilizations are false, which do not civilize the lowest units of any social order, I have written Solaris Farm as my contribution towards the improvement of agriculturists as a class, of the race as a whole; towards the establishment of a truer civilization, organized for the purpose of securing the same degree of progress for the lowest orders of humanity, which have been or can be attained by the highest. In any social or political fabric, wide differences of wealth, of education, of refinement in its sub-divisions are dangerous, they swiftly lead to the introduction of caste. Caste is the dry rot, which, when once established, will surely destroy all progress, all vitality, by slowly eating away the social, industrial and political life of the nation.

In preparing this book for the press, I wish to acknowledge my obligations to the following authors, for much valuable information and inspiration: To Elmer Gates, the discoverer of new domains in Psychology, the inventor and discoverer of the art of Mentation, the founder of the Elmer Gates Laboratory, at Chevy Chase, Maryland: To Henry George, the author of "Progress and Poverty:" To Edward Bellamy, the author of "Equality," and "Looking Backward:" And lastly to that greatest of living

PREFACE.

Frenchmen, M. Godin, the author of "Social Solutions," and the founder of the "Familistere," with its famous industrial enterprise, located at the city of Guise, France; the grandest co-operative success of the age!

A last word to my readers: Do you wish to join forces with the humanitarians? If so, always strive so to educate the people, that they may fully understand the true object and purpose of human life; and the necessity for the upbuilding of social, industrial and political institutions, in harmony with the demands of that purpose. This will require unselfish, persistent, co-operative effort and thought. In no other way, can you so greatly aid the cause of progress.

MILAN C. EDSON.

No. 1728 N. J. Ave., N. W.
Washington, D. C., Sept. 1st, 1900.

CONTENTS.

CONTENTS.

SOLARIS FARM.

A STORY OF THE TWENTIETH CENTURY.

CHAPTER I.

A FARMER'S SON WITH PROGRESSIVE TENDENCIES.

ONE bright summer afternoon, near the close of the month of August, 1905, two young college chums, Fillmore Flagg and George Gaylord, just met after a long separation, were seated on a rustic bench near a well-appointed mountain hotel. The superb view before them was well worthy of their half-hour's silent admiration. Full one thousand foot above the sea stands "Hotel Mount Meenahga" in the heart of the "Shawangunks," a mountain range in the state of New York, famed for its scenic beauty, cool dry air, pure water and commanding elevation. Looking northward a most charming landscape presents itself, a wonderful group of mountain ranges, stretching for seventy-five miles from near the Delaware Water-gap eastward to and including the Alpine peaks of the famous Catskills. Within this lovely semicircle lie the highlands of Ulster, Sullivan and Orange, lifted like seats in some vast amphitheater, tier above tier, while nearer a beautiful mingling of villages and hamlets, broad fields, green woods and silvery water-courses, constitutes a picture of enchanting beauty—a picture constantly changed, shaded and intensified by broad patches of moving shadow and sunlight from a great fleet of fleecy clouds sailing so swiftly, so silently and so majestically across the summer sky.

"How exquisitely beautiful!" murmured Fillmore Flagg, "I wish I had my camera that I might make it captive, carry it hence and keep it, a rare token of beauty, a source of joy forever."

At this point, a brief description of the young men will serve by way of a further introduction.

Fillmore Flagg was fully six feet in height, though his compact, well-rounded figure made him seem less tall; his straight, muscular limbs were in harmony with his deep chest and symmetrical shoulders. His rather large but beautifully turned neck and throat rose straight from the spinal column, firmly supporting a noble head, everywhere evenly and smoothly developed. His thick, soft brown hair, worn rather short, was inclined to curl, giving to the outlines of the head a still more heroic size. His forehead was large, full, dome shaped and remarkably smooth; the brows, finely penciled and well arched, were matched in color and slenderness by a short moustache which seemed a shade or two darker than the hair. His eyes were large, very expressive, of a soft dark brown, bright and flashing with emotion, full of pensive light when partially shaded by their thick silken lashes; his smiling glance possessed a curiously fascinating magnetic charm. The attractiveness of the entire face and neck was intensified by the wonderful marble-like smoothness of skin which accompanies that rare, pale olive tint of complexion. A soft Alpine hat and a neat business suit of dark clothing completes this picture of the personal appearance of Fillmore Flagg. Later on we shall learn to know him better by his genial temperament, mental and moral characteristics.

George Gaylord was above medium height, slender and pale, slightly inclined to stoop; wore glasses, and a thick black moustache which entirely concealed his thin lips. His heavy growth of long, coal black hair was naturally bent on falling over his high white forehead. His large black eyes were deeply set under heavy dark brows, more square than arched. His straight nose and smoothly shaven chin were set in line with his high square forehead. While both face and figure suggested the student, a tall silk hat and a square cut, closely buttoned black frock coat, stamped him at once as a clerical student.

"Tell me, George," said Fillmore Flagg, "how have you fared since we parted, and what are your ambitions and plans for the future?"

"There is not much to tell you, Fillmore. As you know, when I left college, my mother was a widow with a very limited income, which made it difficult to meet my college expenses. Mother had set her heart on my entering the ministry. Her only brother, a childless widower, and a man of some wealth and great influence in the church affairs of his prosperous New England town, promised his assistance. Behold the result! I have just graduated with fair honors from a prominent theological institute. I am to take charge, this coming November, of a large church and congregation in the manufacturing city where my uncle resides. Uncle George, for whom I was named, is now with my mother visiting friends in New York. They have kindly selected as my future wife, my uncle's favorite niece and prospective heiress to his wealth. When last we met, four years ago, Martha Merritt was a sweet little miss in short dresses; but gave promise, even then, of unfolding into a lovely woman. To tell you the truth, under the circumstances, I am more than half prepared to fall in love with her when we meet again. However ambitious my day dreams in the past may have been, a not unkindly fate has woven the web of destiny for me and fixed my future life work without much effort on my part; and yet I am quite content to have it so. Two weeks ago I left the heat and bustle of the great city for a month's rest in this quiet place. I little dreamed of meeting you here; I need not say I am delighted: I am, thoroughly so. I find you looking your best, yet I can easily perceive you have been hard at work as usual. I do not believe you could possibly keep still and rest, even for one short week, let the inducement to do so be ever so great. And now, my dear Fillmore, since I have, so to speak, brought myself up to date for your benefit, may I ask for a similar service on your part?"

CHAPTER II.

THE OUTLINES OF A GREAT PROBLEM.

FILLMORE FLAGG, seemingly self absorbed, remained silent for some moments, softly stroking his chin with his strong, shapely hand, his dreamy eyes with far-off vision intent, apparently noting details in the hazy borders of the distant landscape. At last, turning to his friend with a hearty hand clasp he said: "George Gaylord, I congratulate you; your future is bright; you deserve it, your mother deserves it. The fates have been very generous with you. I am glad you are content to accept the good things of life which they bring to you.

"As for myself, my lines of life are cast in swift waters. My environments, in their reaction upon me from within, seem to develop a determined will to wrench from the rocks of destiny by ceaseless and persistent effort, whatever gifts I am to possess or enjoy. Work I must. Obstacles seem only to stimulate my ambition to overcome them. Yet I am passionately fond of the beautiful; poetry, music and art in all the loveliness of its varied forms; they affect me profoundly. This poetic side of my nature I inherit from my dear, devoted mother—my highest ideal of all that is good, lovely and angelic in woman. Sadly and often have I missed her loving tenderness, her watchful care, her beautiful smile. The shadowy Angel of Death claimed her and bore her from my sight when I was but four years old. Young as I was at that time, this beautiful world has never seemed quite so bright to me since.

"My father, Fayette Flagg, was a noble man of sterling worth. He belonged to a class of thrifty, hard-working, pioneer farmers, on the broad, fertile prairies of the state of Nebraska. Until the death of my mother he was happy and prosperous, hopeful, helpful and brave. After that great blow came to him, he recovered slowly, as from a long, severe illness and never again was quite so courageous and strong, or as hopeful as before.

"With the advent of the last decade of the nineteenth century a feeling of foreboding unrest seemed to brood over the western farmer: blight and drouth destroyed his best crops just when they seemed to promise most; farm stock had to be reduced. The good years were few, the bad years were many. The great strain of carrying a large outfit of expensive agricultural machinery which on a small farm could be used with profit only from ten to forty days in the year, began to be felt. The debts, incurred by the purchase of the machinery, were growing steadily larger. With each renewal of the mortgage on the farm, came the demand for a bonus and a higher rate of interest. Meanwhile the price of land and of all farm products kept on falling, falling steadily year after year. Only taxes and freight rates from farm to market kept up. High rates of interest and of freight swallowed up everything and seemed to accelerate the terrible shrinkage of values. My father found, to his amazement, that his farm was now mortgaged for more than it would sell for under the hammer. He gave up the struggle in despair. The savings of a lifetime, his health, strength and courage all exhausted; his homestead and farm sold from under him; he lost all hope and in a few short weeks died, a broken-hearted man. I went to him a few months before the end: I tried all in my power to save him, but alas! I could do nothing but bury his body beside that of my mother and come away, filled with the determination of solving the most difficult problem of a lifetime—a problem that lies at the very foundation of the permanency of this republic. 'How to keep the farm lands of America in the hands of the native farmers of this and the coming generations? How to help them to help themselves?' The decree has gone forth. The small farm and farmer must go. They are doomed. A great wave of land monopoly, rolled up by a large class of very shrewd, far-seeing capitalists, is even now sweeping across the continent. Seventy-five years hence only a pauperized peasantry of ignorant farm laborers, bound to the soil as hopelessly as the slave to the master, will coin their lives of ceaseless, unrequited toil to swell the rent roll of the non-resident landowner, who, as lord of the domain, through his heartless agent, will exact his tribute to the uttermost farthing. Must the sons and daughters of the farms of this republic come to the bitter heritage of such a life? Surely! We have already seen the beginning of the end! The sad case of my father can be duplicated a hundred times or more in almost every county of our western states. States that are incalculably rich in their magnificent domain of broad acres of the most fertile land the sun ever shone upon; capable, when permanently placed in the hands of a properly equipped, scientifically educated class of people, of producing the food supply of the world: but under the blight of the monopoly system, history will repeat itself. Our agricultural interests will languish and wither; dependent manufactures, and all branches of exchange and commerce, must, in time, follow. What then will happen to society? To government of both state and nation? In the face of this appalling situation, how stupendous the problem! By what effort can a great counter tidal-wave be set in motion upon whose crest the salt and salvation of the republic, the sons and daughters of American farms, may be carried safely to the permanent heritage of the soil they till? As in the past, so in the future must we look to them for our true reformers, leaders, thinkers and statesmen. They are endowed by birth, by constant association in youth with soil and sunlight, fields and grass, green meadows and mossy brooks and, best of all, doubly endowed by the inbreathing of ozone laden breezes from mountain and forest, with that rare combination of nerve, moral, mental and physical stamina, courage and patriotism which is necessary to preserve this republic and to keep it, ever and always, a model of progressive excellence for all the nations of the earth. This means the embodiment by them of more and better mind, that they may do better, wiser and more dominant thinking; be able to comprehend the sum of human knowledge to such an extent that they may add to it; to so understand their lives, and their relations to the Universe around them, that they may become masters of themselves and their environments—a law unto themselves—fitting them for a perfect citizenship of a perfected republic. This most desirable of all accomplishments, requires better surroundings, more leisure and opportunity for self-improvement, more money, shorter hours of more remunerative labor—labor transformed from a hated drudgery to a desirable occupation. Behold, friend Gaylord, you have before you the outlines of the problem. Can you suggest anything towards its solution?"

"I can suggest nothing," said George Gaylord; "You have stated the case with the clearness and eloquence of a Henry George. If what you say is true, the problem is a very serious one. But are you quite sure the facts will fully warrant your conclusions? If so, what are your plans and what have you been doing towards working out this puzzling question?"

"Oh yes!" said Fillmore Flagg, "I am very sure of my position. The more I study the question, the firmer my conviction that I have understated the case instead of overstating it. I am studying the agricultural question from every possible standpoint and I propose to make it a life work. Every branch of science may aid me; I must master at least a portion of each. Since we left college I have become fairly proficient in surveying and civil engineering; have devoted considerable time to photography; I am classed as a skilled electrician; I have thoroughly mastered agricultural chemistry and several of the more important branches of that interesting and most wonderful science. As you know, I am very fond of mechanics and of all kinds of machinery. I could not rest until I had gained a practical knowledge of all kinds of tools and learned how to repair or construct most kinds of machinery. Two months ago I completed a general course of study at the Philadelphia School of Industrial Art, which, for the especial work I have in view, I consider by far the most beneficial and practicable of all my acquirements. I am now resting, cogitating and waiting for the golden opportunity which, sooner or later, must come, to enable me to commence my work."

CHAPTER III.

AN ADVERTISEMENT INTRODUCES THE HEROINE.

"By the way, I have something to show you. I clipped this advertisement from a leading New York daily paper this morning, and have read it carefully many times. Somehow, I have an abiding conviction that it will lead me to the high road, on the way towards the successful solution of my problem. I am going to apply in person."

Full of curiosity, George Gaylord took the clipping and slowly read aloud:

"WANTED: A skilled mechanic, qualified to act in the capacity of landscape gardener and agricultural chemist. Applicant must be a strong, healthy young man, of good habits, pleasing address; with a general knowledge of business methods, and an excel lent moral character. Qualifications must be well attested by recommendations from reliable parties. A graduate of the Philadelphia School of Industrial Art is preferred. Salary liberal. Apply in person at the office of

BITTERWOOD & BARNARD, Atty's.,
Atlantic Building, Washington, D. C."

"This is curious! It seems to point directly to you, Fillmore. I do wonder in what peculiar capacity you are to act, and who your real employer is to be? I shall be full of unsatisfied curiosity until I know the sequel."

At this moment George Gaylord was suddenly interrupted by an an unlooked-for gust of wind whirling around the shoulders of the big rock standing above and behind them. The fluttering paper slipped from his fingers and went sailing away over the tree tops, down the mountain side, with that erratic up and down, eddying motion peculiar to run away, fly away papers. In an instant both young men were upon their feet, intently watching the uncertain flight of the clipping. A few moments later it fell to the ground, just at the feet of two ladies who, with heads protected from the sun by large parasols, were slowly walking around the bend of the broad, well kept road, winding down the mountain side. The younger of the two ladies picked up the advertisement, hurriedly scanned it, and then raised her eyes to discover the two young men as probable owners of the truant paper.

"Ah!" said George Gaylord, "I recognize those

people. It is Miss Fenwick and her travelling companion. Come along Fillmore, let us join them at once and claim your lost clipping. The opportunity for an introduction to two very interesting ladies, who are among the most noted guests of the hotel, is too good to be lost."

Accordingly they hurried down the steep path that joined the road near where the ladies were still waiting, at a point full three hundred feet below.

Approaching, with hats in hand, George Gaylord said: "Allow me, Miss Fenwick, to introduce to you my friend and college chum, Fillmore Flagg: for a peculiar purpose of his own he wishes to regain possession of that flighty paper which, fortunately for him, the prank playing wind carried to your feet but a moment ago."

With a slight inclination of her queenly head, she turned with a dazzling smile to meet the inquiring glance of Fillmore Flagg. In a clear musical voice, full of thrilling cadence and power, she said: "Mr. Flagg, if you are particularly interested in this paper, I am very sure I am quite happy to meet you, and take pleasure in returning it to you now; I trust that we may have the opportunity of becoming better acquainted before you leave these lovely mountains." Turning to her companion she continued: "Permit me, gentlemen, to introduce my friend and companion, Mrs. Bainbridge; Mr. George Gaylord, who is just entering the ministry, and his college friend, Mr. Fillmore Flagg."

Mrs. Bainbridge responded with a pleasant smile. She was a tall, well formed, well preserved woman of forty; full of a quiet dignity, with an air of refinement that fitted her like a garment. Her heavy dark hair, coiled high on her shapely head, was just slightly silvered with gray and seemed to be a fitting foil to her large melancholy black eyes—eyes that from their slumbering depths seemed to impress the beholder with suggestions of some mysterious power, gleaming messages, like beacon flashes, from her inner life. With her becoming dress of rich, dark cloth, gloves and parasol to match, she looked the cultured lady to perfection.

Turning her steps up the mountain, Fern Fenwick said: "Gentlemen, as it is near the hour for supper, we had best return to the hotel at once. I think too, by this time the mail from the station must have arrived." Fillmore Flagg was at her side in an instant, choosing the side opposite the parasol, which gave him a clear view of her charming profile. George Gaylord and Mrs. Bainbridge followed a little more slowly. The conversation soon became animated.

While they are thus occupied let us try to get a more complete picture of Miss Fern Fenwick. Her round, exquisitely proportioned figure was of medium height, straight as an arrow, full of grace with every movement. Her quick, firm, elastic step was Youth personified: a charming maiden, she, of twenty summers. The artistic outlines of her plump arms and shoulders, beautifully modelled bust, throat and neck, so admirably proportioned, would have satisfied the most carping critic; poet or painter, he would have pronounced them a dream of perfect symmetry. Her queenly shaped head, so gracefully poised, like a clear cut cameo, was a poem of intellectual development on lines of rarest beauty. Her thick, glossy hair of dark chestnut brown, fine as spun silk and inclined to a wavy crimp, was artistically coiled in a most becoming style; small ears of perfect shape, and transparently pink, were set close to the head. The curve of the brow, in perfect line with the pleasing oval of both cheek and chin; a Grecian nose and cherub mouth completed the perfect contour of a face and head of marvellous beauty —a beauty made more brilliant by large, lustrous eyes of blended sapphire and amethyst, flashing jewels of deep violet blue, so clearly expressing the varying emotions by their ever changing tints of sparkling light. Her dress, a close fitting gown of rich, soft, silver gray material, was stylishly made, with a narrow line of lovely lace at the throat; perfect fitting gloves of the same shade of gray, with a parasol to match, completed a costume that seemed to bring out and intensify a most charming complexion of pale pink and white, faultlessly smooth and transparently pure: at once indicative and prophetic of a strong vital temperament, perfect mental and physical health; pure, highly cultured mind and a wealth of personal magnetism—that silent charm of mysterious potency—pervading and surrounding her like the perfume of sweet flowers, winning the unsought admiration, friendship and fidelity of all who came within the radiance of her powerful magnetic aura. All this, and more, Fillmore Flagg perceived and felt. He walked and talked as one in a dream. Never before had he met so fair a vision of female loveliness, with grace so winning, gestures so perfect and voice so musical. His heart, overflowing with a new ecstatic emotion, paid silent homage to this queenly creature. He was lost in admiration. Swallowed up and absorbed by the first incoming wave of a great love. He was lifted out of himself, above and beyond all gross things of earth, into a heaven of pure delight. His better nature was thrilled and profoundly moved. He felt that in the presence of this pure, angelic woman he could never again do an unworthy act. A life work, up to the standard of his highest ideal, was a tribute of devotion he would willingly lay at her feet.

All too soon for Fillmore Flagg the moments flew by. Almost before he was aware of it they were ascending the steps of the hotel. Pausing on the broad veranda for a moment before separating, Fern Fenwick said: "Gentlemen, Mrs. Bainbridge and myself have planned for a carriage drive to-morrow to Sam's Point. We have two seats in our conveyance at your disposal and would be delighted to have you accompany us. May we hope that you both can come with us?"

Fillmore Flagg and George Gaylord both eagerly accepted the invitation, the ladies passed on to their rooms, while the young men turned their steps once more to the rustic bench to enjoy the magnificent sunset view of the landscape they had so much admired earlier in the day.

CHAPTER IV.

THE STORY OF A STONE AND WHAT CAME AFTER.

SAM'S POINT, the crowning backbone of the highest mountain in the Shawangunk range, bends away from the general course of its fellows apparently for the especial purpose of giving the mountain climber, by its isolation, a commanding view in almost every direction except to the north-east. For miles in extent the flat, rocky top of this crown forms a promenade of magnificent proportions up amid the clouds.

In shape it is a long, slender triangle, about three miles from its base westward to the point where its highest altitude is reached, two thousand three hundred and forty feet above tide-water. Cradled in its rocky bosom, near the base of the triangle, lies a crystal lake—one hundred and fifty acres of sparkling water. At this point the promenade is fully three-fourths of a mile wide, gradually narrowing to a width of less than one hundred feet at the extreme point. The long battlemented sides of this lofty triangle, like some mighty fortress, grim and frowning, are protected and supported by perpendicular cliffs of black rock, rising like some bastioned wall of terrifying proportions, two hundred feet above the shoulder of the mountain. In a sheltered nook, near the point, about five hundred feet below the base of the cliffs, stands the Sam's Point Hotel, scarcely more than a cottage in size. Here Fern Fenwick's party left the carriage. Taking the narrow, zig-zag pathway that led to the cliffs and often pausing to admire the immensity and grandeur of the black rock palisades towering so far above them, they soon found themselves under the nose of the point of rocks. Entering the crevice in the cliffs known as "The Chimney Stairway," they commenced the steep and toilsome climb to the summit; Fillmore Flagg taking the lead and assisting Miss Fenwick, George Gaylord performing the same service for Mrs. Bainbridge; fifteen minutes later they stood, almost breathless, upon the summit, the blue sky all about them, a precipice on either hand where shimmering, giddy space seemed to yawn so frightfully near. Meanwhile a strong, buffeting wind tugged at ribbons and capes, hats and bonnets, so furiously that walking was hazardous; it gave one such an uneasy sensation of giddiness and unstable equilibrium generally, that the temptation to fly over the edge of the cliff was hard to resist. A huge egg-shaped boulder, twenty-five feet in height and as large as a house, poised rather unsteadily on its rounded base, was quite near and gave promise of protection from the violence of the wind. With one accord our party scrambled towards it, the ladies clinging tightly to their escorts with one hand, a firm grip on hat or bonnet with the other. Thus sheltered, and more at ease, they slowly drank in the glorious vision which greeted the eye on every hand. Looking down as from a balloon, at the foot of the mountain, on the north side, the eye was charmed by the length and beauty of the Rondout Valley, through which ran the Delaware and Hudson Canal, and the Rondout River. For miles on either side of canal and river the valley was made more lovely by its checkered farms and gleaming white villages. Directly at the foot of the mountain on the south side, the broader valley of the Wallkill presented an equally beautiful and diversified picture of farm, hamlet and village. Beyond these, in every direction save to the north-east, vast stretches of country lay spread out like a map; the mountains far and near, so dwarfed as to give to the surface the appearance of billowy plains, almost level where they approached the edge of the horizon. The wonderful extent and scope of the view was bounded by the line of the horizon, at least one hundred miles distant. Three-fourths of this sweeping circle responded to the unaided vision, disclosing the blue hills and hazy mountain peaks located in five states: New York, Pennsylvania, New Jersey, Connecticut and Massachusetts, altogether presenting in its immensity a landscape as variegated and charming as it was wondrously beautiful and attractive—a marvellous picture of indescribable loveliness never to be forgotten.

"How inspiringly magnificent!" said Fillmore Flagg: "All the sublimity of my nature is satisfied."

"And I," said Fern Fenwick, "am too profoundly impressed to talk. I would that I could spend hours here in silent admiration."

"I think," said Mrs. Bainbridge, "that we would better move further back on the rocky summit where doubtless, sheltered seats may be found, then we can all enjoy this most wonderful of views at our leisure and with some degree of comfort."

"Yes," said George Gaylord, "that will be ever so much nicer."

"Stop a moment," said Fern Fenwick, who for some moments had been examining the huge boulder which sheltered them, "Have you noticed the curious formation of this immense stone? How many hundreds of tons it may weigh, I hardly dare guess. Geologically speaking, it is a 'stranger rock,' not in any way related to the rocks of this mountain, nor of the mountains near here. It is a mammoth conglomerate of such an interestingly curious compound and of such flinty hardness. At the time of its formation enormous pressure, coupled with the most intense heat, must have molded this strange mass together. Coarse and fine gravel, smooth, round pebbles, from the size of a pigeon's egg to that of a two-hundred-pound boulder, are all jumbled together in great confusion, and so firmly cemented in this immense globular mass of that peculiar, tenacious clay of greenish gray color, which forms so large a part of the drift formation, and which is so widely distributed over the face of our globe—that strange, unaccountable, isolated and unrelated formation, which still remains an unsolved puzzle by our best geologists. I wish you to observe the long sides of this strange rock, especially where the exposed sides of the pebbles have been worn down smooth and even with the clay—how they are marked and striated by shallow grooves, all running in one direction as straight as though graven by rule. Is it possible that any freak or flood of the glacial period could have floated this huge rock to its resting place on the very summit of this high mountain, almost two thousand five hundred feet above the level of the sea? Oh! tell me, ye listening mortals, or ye winged winds that blow and pull my ribbons so! whence came this stranger rock? how formed? and how were its smooth, worn sides so systematically engraved?"

Fern Fenwick closed her series of queries with a gradually rising pitch and inflection in the ringing tones of her clear, musical voice. With figure erect, eyes flashing, cheeks glowing and hands uplifted, she seemed the personification of some priestess of science. Fillmore Flagg and George Gaylord gazed at her with the admiration of amazement. Mrs. Bainbridge exclaimed:

"Why Fern Fenwick! How you do go on with such nonsense, to be sure. No doubt these gentlemen, from this time forward, will look at you as some scientific freak or geological professor of the female persuasion, but recently escaped from the walls of some famous college!"

"Mrs. Bainbridge," said Fillmore Flagg, "of course we understand that you were joking in what you said just now: that you really admire the terse, clear, and wonderfully complete description of this strange rock by Miss Fenwick, quite as much as we do." Turning to Fern Fenwick, he continued: "I believe, Miss Fenwick, that I can throw some light on the puzzling questions you have so poetically propounded."

"Pray do tell us, Mr. Flagg," said Fern Fenwick; 'I can't remember when I was so excited with interest on any subject before."

"Very well," said Fillmore Flagg: "That curiously able and intellectual man, Mr. Ignatius Donnelly, in his very interesting book called 'Ragnarok,' or The Age of Fire and Gravel, puts forth a most remarkable theory regarding the drift formation, to the truth of which this huge rock seems to bear witness. The theory, briefly stated, is as follows: A great many ages ago, when this globe of ours was still in the period of cataclysms, rolling through space around the sun, it came in contact with a portion of the end of the tail of some enormous comet, sweeping through the universe on its erratic course. This great boulder is a sample of the component parts of that firey tail, which smote the exposed face of the earth so terribly with the drift deposit at that time of dire disaster. The age of fire and gravel, surely! This curious clay, now of such flinty hardness, was at one time the exceedingly fine dust of the comet, cohering, collecting and embedding its mixture of pebbles and gravel by the heat and pressure of the friction caused by its incalculably swift passage through space for periods of uncounted ages. Remember that the heat of all drift material in the tail of the comet was greatly intensified by the explosion of accompanying gases as they came in contact with the atmosphere of our earth, All inflammable material on the face of the globe, which was exposed at the time of its passage through the tail of the comet, was burned up: both earth and sky were on fire! Fortunately our flying globe made a quick passage, thus it happened that large portions of its unexposed surface wholly escaped this terrible downpour of fire and gravel, and the absence of all drift deposit on these places is logically accounted for. The atmosphere, so heated during that awful period, drank up the waters of the earth—then came the floods, as the waters fell again. Then followed the reaction period of extreme cold, snow and ice—the glacial period. This particular rock, while following in the train of its parent comet, though lagging many thousands of miles behind, still, being so very large, moved with accelerated speed towards the comet's head, passing on its way countless millions of smaller particles, whose cutting edges scored these grooves. On entering the earth's atmosphere, on account of its great size, this boulder, through the law of attraction, quickly moved to the outermost fringe of the comet's tail nearest the earth, therefore was the first to alight on the top of this mountain, far away from all smaller drift material.

"I hope, Miss Fenwick, that my brief and rather speculative answers to your questions, reasoning as I did, from Mr. Donnelly's point of view, may prove at least in a measure satisfactory."

"Thank you, Mr. Flagg," said Fern Fenwick, "your answers to my questions have all been very ingenious: equally interesting and satisfactory, especially as to how this mammoth conglomerate came by its grooved lines and, later how it managed to find a resting place on this mountain top, so far from its kind. Mr. Donnelly's theory of accounting for the widely scattered deposits of the drift formation is the most reasonable and logical of anything I have ever read or heard. Doubtless, in course of time, it may be proven the only true one. I see Mr. Gaylord and Mrs. Bainbridge are becoming weary of all this talk about rocks: let us move further back from the point in search of more sheltered and comfortable seats."

Accordingly they chose the central path and were soon seated, enjoying the changed landscape from a new point of view. However, Mr. Gaylord was not yet satisfied and soon proposed a walk to the lake. Mrs. Bainbridge was willing but Miss Fenwick had walked enough for one day. A quiet enjoyment of her lofty outlook was what she now most desired.

"Very well, Fern," said Mrs. Bainbridge, "Mr. Gaylord will accompany me to the lake and we will bring back for lunch some of those very large, delicious blueberries, which Mr. Gaylord assures me are growing so abundantly around the shores of the lake. You and Mr. Flagg shall remain here with the lunch baskets."

This plan was agreed to, and very soon Mrs. Bainbridge and her escort had disappeared on their way to the lake. To Fillmore Flagg it seemed a long time that Fern Fenwick had been sitting so quietly, apparently absorbed in admiring the billowy miles of landscape unrolled so far to the southward. In reality, each was thinking of the other.

"Mr. Flagg," said Fern Fenwick slowly, "will you pardon me for asking you some very abrupt questions, or what may seem such when considering our brief acquaintence?"

"Certainly," said Fillmore Flagg, "I hope my replies this time may prove as satisfactory as those I gave in regard to the rock. The pardon you crave is granted in advance. Pray proceed."

"Tell me, Mr. Flagg, why are you so much interested in that advertisement which came to me so unceremoniously yesterday? And again, tell me why you are so moved and determined to better the conditions of farm life? I suppose you know that I have wealth and leisure at my disposal; it may prove that I can be of great assistance to you. This is my excuse for asking you for more details in regard to your personal plans."

With a heart filled with hope, Fillmore Flagg began the recital of the story he had given to George Gaylord on the terrace bench. With frequent glances of encouragement from Fern Fenwick, his inspiration and eloquence grew upon him. He gave a masterly statement of the work, his preparation, hopes and plans. Delighted beyond measure with the undisguised appreciation and approval of this charming woman, whose very destiny in the vista of a coming future, seemed to him to be linked in some mysterious manner with the success of his most cherished ambitions, he cleverly enlarged and perfected the original statement. As he concluded, Fern Fenwick rose to her feet with hands extended, her face glowing with interested enthusiasm, saying:

"Mr. Flagg, I most heartily congratulate you on the noble life-work you have planned and chosen, I thank you again and again for the valuable facts you have placed so confidingly in my possession, in regard to yourself and your work. Rest assured my interest and assistance henceforth are at your command. You will understand this more clearly when I tell you that Bitterwood & Barnard are my attorneys, and the advertisement which played such an important part in bringing us together here in these mountains, was drawn up by them for my purposes. That it should bring to me a person of your wonderful ability, integrity, skill and knowledge, is an almost unhoped for piece of good fortune. You are the one, of all others, most eminently fitted to help me to a successful solution of my problem, which you have so admirably stated. Hereafter I am your debtor. I hope to prove a not unworthy employer, or, to put it more pleasantly, an interested co-worker. Will you do me the favor of considering yourself as pledged from this moment to take up my work? Go at once to my attorneys in Washington, ask them for a letter of introduction to me, that you may get more complete details of my plans and work, saying not a word of our present acquaintance. I will furnish you with a check on my Washington bankers. with which to defray your expenses. To-morrow, in company with Mrs. Bainbridge, I go to my summer home on the Hudson near Newburgh, where letters will reach me. This is the twenty-eighth of August; on the fifth of September, at noon meet me in the station at Newburgh. Come prepared to devote a week at the least in discussing the scope and plan of our work, devising ways and means etc. I very much desire that you have an interview with my father, I know he will be pleased with you. Do these arrangements suit your convenience? Do they meet your entire approval?"

"I am greatly elated," said Fillmore Flagg, "at this my golden opportunity of commencing what you have so kindly named as 'our' work, under such auspicious circumstances. I thank you, Miss Fenwick, more than words can tell, for your confidence in my integrity and ability, I will do my best to retain that confidence. I am ready to start for Washington to-morrow. I will follow your instructions, and will report to you by letter from that city, and then meet you at Newburgh at the appointed time."

As he finished his reply Fern Fenwick said: "Mr. Flagg, I am very much pleased with your prompt decision in favor of my arrangements. I see our friends returning from the lake, will you help me to spread the lunch?"

With keen appetites they enjoyed the lunch especially the delicious blueberries which George Gaylord and Mrs Bainbridge had brought from the lake. The hours passed quickly; the drive back to the hotel was without mishap or incident: the entire party, on separating, voted it a day of perfect pleasure, Fillmore Flagg and George Gaylord expressing their thanks to the ladies for their kind invitation which had given them such a delightful excursion.

Later, George Gaylord called at the room of his chum for a few moments chat. "Come in," said Fillmore Flagg, "I was just thinking of you. I have made up my mind to go to Washington to-morrow for the purpose of answering that advertisement. How much longer do you propose to remain here?"

"Not more than two weeks," replied George Gaylord. "I understand Miss Fenwick and Mrs. Bainbridge are going away to-morrow. I am likely to have a very quiet time, all by my lone self: I think I must take to bowling for an hour or two each day just to keep up my exercise and kill time. I hope you may be entirely successful in your interview with Bitterwood & Barnard. Remember how much I am interested in this matter, and your promise to let me know the result. By the way, what a perfectly delightful day we have had, thanks to that lucky gust of wind which tore your clipping from my fingers and landed it at Miss Fenwick's dainty feet. What a talented young lady she is, and so handsome too. Her lecture on the mountain top about that stone would have been a credit to any one. I never saw her look such a picture of perfect beauty before. She seemed wonderfully interested in you, Fillmore, especially after your brilliant reply to her series of apparently unanswerable questions. I declare, the profoundness, the ingeniousness, and the boldness of your successful answers filled me with amazement! You fairly surpassed yourself; all the time looking your best, just like a hero. Yet when you looked at Miss Fenwick you seemed just at the point of falling down to worship her. I can't blame you. What a glorious couple you two would make! If it were not for her immense wealth I believe you could win her; any one can see that you have made a very favorable impression. Perhaps you can win her as it is—I wish you all success, you certainly deserve it. Mrs. Bainbridge tells me that at the death of Miss Fenwick's father, some years ago, she became sole heir to his vast fortune; most of it in very rich Alaska gold mines."

"Are you quite sure," said Fillmore Flagg, "that her father is dead?"

"Yes Fillmore, I am quite sure; although it is just possible that I may have misunderstood Mrs. Bain bridge. In my hotel acquaintence with that lady I discover that she is a very intelligent and accomplished person of rare good sense. Splendid company; we seem to get on famously together, I shall miss her very much I am sure. As usual, I am doing all the talking: it is now your turn to say something."

"I think I could," said Fillmore Flagg, "if my

chatterbox friend, George Gaylord, would only give me a chance. Miss Fenwick I regard as the most beautiful and cultured woman I have ever met. I do admire her very much, but the possibility of ever winning her for a wife is, at this time, too remote for me to consider for a moment. I must now pack my trunk and then see the hotel clerk about getting it to the railway station. So good night, George, I will see you again in the morning."

That night Fillmore Flagg could not sleep. The beautiful image of Fern Fenwick was before him the moment he closed his eyes. The events of the past two days, with their crowding memories, kept racing through his mind: he could not think calmly or connectedly. He was in a fever of expectancy regarding the meeting at Newburgh, and the prospect of spending a whole week at Miss Fenwick's cottage on the Hudson. Then and there, no doubt, she would tell him all about herself, her father, her particular work, when and why she became interested in it etc. But what about the father? How could he have an interview with her father, if Mrs. Bainbridge was correct in saying that Mr. Fenwick had been dead for several years? It was a mystery he could not solve. He did not doubt Fern Fenwick for a moment and felt sure she would, at the proper time, make everything plain. How gracious and winning she had been to him; she seemed to bid him to have courage. In spite of her great wealth, and a hundred other obstacles that might exist, he was more and more in love every hour. If proving himself worthy of her confidence in every way would win her love, surely then, he would win it. With this determination fixed in his mind he fell asleep.

In her room that night, as Fern Fenwick brushed

her hair and prepared herself for rest, she often paused to ponder over her strange meeting with Fillmore Flagg; thinking what a fine, manly looking fellow he was, and how well he could talk; how thoroughly equipped he was to take up the question of improving farm life, the lives of farmers and their families—the question of all questions for her. Surely, Mr. Flagg bore the stamp of destiny! He was the man of all men to make her work a complete success. How fortunate she was to secure his valuable services. How strange, that after a brief acquaintance of only two days, she should have such perfect confidence in a comparative sttranger. Yet, she did not doubt his integrity; she knew he was loyalty itself; she intuitively felt that she could trust him implicitly—he would never betray her interests under any circumstances. She knew from his every look, tone and gesture that he admired her intensely, devotedly. Her own feelings, she did not care to analyze. With a sigh, more of pleasure than weariness, she composed herself for the night and was soon lost in sleep.

CHAPTER V.

FAIRY FERN COTTAGE.

ONE WEEK has passed since the events narrated in the previous chapter. At Cornwall on the Hudson, on a West Shore train speeding north, we find Fillmore Flagg; his mission at Washington successfully accomplished, the letter of introduction from Bitterwood & Barnard secured. In another short hour he will be at Newburgh. Will the lovely face of Fern Fenwick be the first to greet him? As the moments fly by, his heart beats faster. He feels the surging tide of his all-absorbing love for this beautiful woman, thrilling and permeating his entire being. He tries to be calm, to think what he ought to say that would be fitting and appropriate; he knows his eyes are blazing and his cheeks glowing with an unwonted fire, still his thoughts refuse to flow into the satisfying forms of speech he most desires to use at the coming meeting, which seems to him to be the marking of a great crisis in his life. Ah! There is the whistle sounding! The speed of the train is checked as it approaches the station. He steps on to the platform while the train is still moving. He beholds many upturned faces in the surging crowd between him and the doorway of the ladies' waiting room, but Miss Fenwick he cannot see. Will he ever reach that room? Has anything happened to her? A great fear contracts his heart, he fancies he fairly staggers as he enters the door. In an instant he is suffused with a great joy. By the window, awaiting his approach, stands Fern Fenwick, the perfect picture of cool, contented loveliness. She extends her hand and greets him with a firm clasp of hearty welcome, and a second edition of that dazzling smile, so becoming to her, so bewitching to him.

"How do you do, Mr. Flagg? I believe your train must be late. How well you are looking, in spite of the heat and the dust! We will have your baggage secured as soon as possible and placed in the carriage, then we will drive to the cottage in time for lunch."

"Thank you Miss Fenwick, I am delighted to see you looking so well. My journey from Washington has been a very pleasant one; I have enjoyed it and have not suffered from the heat."

The carriage now came up, they stepped in and commenced the beautiful drive of one and one-half miles to "Fairy Fern Cottage," which was charmingly located on the summit of these famously terraced hills. Hills that have been historic since the revolutionary days of General Washington, when their slopes were white with the tents of his soldiers. As they approached the cottage, the artistic eye of Fillmore Flagg noted with pleasure the broad expanse of spacious lawn, gently sloping down to the road. Half-moon-shaped, it presented for his admiration five acres of smoothly shaven, velvety green. For one-eighth of a mile, the entire width of the lawn and cottage grounds, a low wall of ornamental cut stone separated the lawn from the road and formed the straight line of the half-moon. From the gates at either end of the wall a broad, beautifully kept driveway swept around the semicircle of the lawn, passing just in front of the cottage at the center of the deep bay of the half-moon. On each side of the driveway the greensward was beautified by alternating star and diamond-shaped plots of geraniums, roses, gladioluses, canna and nasturtions. Sitting close to the outer edge of the drive, about ten feet apart, commencing at the corners of the porch on either side, were rows of potted palms extending around the curve, one hundred and fifty feet each way—the palms gradually growing smaller as the distance from the cottage became greater. The effect was beautifully unique and suggestively semi-tropical. The cottage and lawn was embayed by a crowning crescent of choice foliage and shade trees; the thin horns of the crescent terminated at the gateways in low gray stone towers. From these points the horns gradually grew broader and the shrubbery rose higher. First the rhododendrons mixed with clumps of hollyhocks, next flowering almonds, roses, spireas and syringas; then came the drooping long leaf sugar pines, with an artistic mingling of slender limbed graceful silver birches: farther back were the taller firs and spruces, interspersed with thick clumps of small copper beeches, extending to and joining at the back of the cottage, the dense forest of tall, straight bodied elms, oaks and maples which partly hid and shaded the stables and the kitchen portion of the cottage.

The cottage itself was built of gray stone; with thick walls and large, low, deep seated windows. It was two stories in height, with three square towers rising twenty feet higher. The central tower was larger, and gave space within its walls for one grand room of magnificent proportions, thirty feet square and with a fifteen foot ceiling. The general effect of the cottage, lawn, and crescent background of foliage and forest, was as novel as it was beautiful. As the carriage entered the farther gateway, Fillmore Flagg was surprised and delighted:

"How perfectly exquisite!" he exclaimed: "A real gem! A romantic scene from fairyland! Rightly named 'Fairy Fern Cottage!' It is a fitting home for Fern Fenwick."

"Thank you, Mr. Flagg," said Fern Fenwick as they stepped from the carriage to the porch: "I appreciate your praise of my cottage home. I love it, I am proud of it, I give you a hearty welcome to its halls. May your memories of it prove always pleasant. Let us enter. During your stay you are to occupy the front room on the second floor, the one under the right hand tower. I think you will find the view from the windows very pleasing and attractive. The luncheon bell will sound in just half an hour."

In the dining room Fillmore Flagg found Mrs. Bainbridge who greeted him very cordially. She sat at the left of Fern Fenwick, who was at the head of the table. The table itself was oval shaped, very large, seemingly of rich, solid mahogany; the china and silver were elegant and artistic. The center piece was a large silver tray filled with a wonderful collection of rare ferns. Around it a ring of cut glass bouquet holders, filled with spikes of flaming gladioluses, formed a most effective border.

"You are to sit here at my right, Mr. Flagg," said Fern Fenwick.

As Fillmore Flagg took the proffered seat, he thought her a most charming hostess, admirably fitted to preside over this exquisitely decorated table. He looked in vain for her father; finally concluding that Mr. Fenwick must be a confirmed invalid, confined to his room. Luncheon over, Fern Fenwick invited Fillmore Flagg to her study to consider the business of the work before them. Her study proved to be the large square room in the central tower, which was so generously lighted by its eight large windows. The furniture was of carved oak; the carpet and hangings, rich and heavy, were of a pale lilac tint, which gave an air of peaceful quiet and harmony to the room. From the front window, looking eastward, a long stretch of the beautiful Hudson could be seen at one sweeping glance. In the south east corner of the room stood Fern Fenwick's desk, a large one with a roll top. At the right of the desk, on an easel against the wall, was a very fine, life size crayon portrait of a noble looking man of sixty winters or more. The massive forehead was both broad and high and very smooth. The eyes were wide apart, large and expressive, the full beard, thick and fine; the hair, abundant and wavy. Both hair and beard were evenly tinged with gray. The body was large, erect and well proportioned—it fittingly matched the noble head. The portrait impressed one as being life-like and full of character. Close beside the easel was a large arm chair, upholstered with stuffed leather, a grayish brown. Lying across the arms of the chair was a large, peculiarly shaped trumpet of aluminum, ornamented with a heavy cord and tassel of gray silk.

"Mr. Flagg," said Fern Fenwick, "this is my private workroom; here I am undisturbed and not at home to callers. This is my desk. Here you see my father's portrait: this is his favorite chair. Will you be seated in the smaller chair near it? I will sit in the chair at my desk."

"Pardon me, Miss Fenwick," said Fillmore Flagg, "Up to this time I had thought of you as living here with your father: I now perceive, from the way you speak of his portrait and of his favorite chair, that he must be dead. Please correct me if I am wrong in my conclusions."

"I will explain the situation in a very few words," said Fern Fenwick.

"In the eyes of the world I am an orphan, my father and mother having both passed from this to the land of spirit. The world, in its blind ignorance, calls them dead. To me, thanks to my mediumship, and to the mighty truth of spirit communion, they are still conscious, living, loving parents. Every day, here in this room, they come to me and through the trumpet there, speak to me as naturally, as fluently and as lovingly as ever. I feel and realize their constant watchfulness and loving care. In times of need their advice never fails, always proving as wise as it is unerring. They never for a moment allow me to realize that I am an orphan in any sense of the word. The word Death has no terrors for me: I realize that for them it means simply a happy transition to a higher life, filled with broader and brighter possibilities; and, blessed truth! that they are permitted to come to me when I need them. I sometimes shudder when I think what might have happened to me if I had not been born and bred a spiritualist and a medium. However, we will speak of these things more at length later on. At this time, under my father's guidance and with your assistance, I am to cary out and complete his plans for the improvement of farm life on lines quite in harmony with your ideas. I know he approves of you and of your work, and has confidence in your integrity and ability. At the proper time he will speak to you personally through the trumpet. Let us now consider another matter pertinent at this time.

"In order that you may thoroughly understand the situation that surrounds and affects our work, it will be necessary for me to tell you the story of my life, and with it the story of the life of my father."

CHAPTER VI.

FENNIMORE FENWICK.

"ON A pioneer farm in northwestern Iowa, with a broad expanse of beautiful prairie on every side, far from town or village, lived my grandfather, George Fenwick. On this farm in October, 1840, my father, Fennimore Fenwick, was born. Of a family of nine children, five boys and four girls, he was the fifth, two of the brothers and two of the sisters being older. Closely associated as a healthy, harmonious family of children, they grew up surrounded by the conditions of an isolated farm life, so general in the widely scattered settlements of those early days, with only now and then rare chances for a little schooling of the most primitive character. However, they shared with each other their joys and sorrows, their plays and privations; always forbearing and patient, kind and affectionate, light-hearted, sympathetic and helpful, they did much to develop that broad, loving, genial nature which made my father kin to all mankind. So just and true! So nobly unselfish! A signal illustration of the great blessing which Nature's beneficent law of compensation brings to large families.

"Passing on to September, 1865, at the close of the war of the rebellion, we find the large family, so long and harmoniously united, now separated and widely scattered. Grandfather and grandmother Fenwick both died during the closing year of the war. With the exception of my father, the brothers and sisters were all married and settled on farms of their own: some in Iowa, one in Missouri, two in

Kansas, and two in Minnesota. The homestead was divided between the two younger brothers. All of the brothers served as soldiers, good and true, during the war; the two younger only one year each. My father, more fortunate than the others, by his bravery and soldierly excellence won a commission, and came home the captain of his company.

"From this point forward we will follow my father's career as he makes a pathway in life for himself.

"From 1865 to 1871 he devoted his time and his savings to hard study in the best of schools, finishing a master of his profession—a mining engineer and expert in assaying and metallurgy. From 1871 to 1882 he was general manager of a wealthy mining company in Colorado at a large salary, making a name for himself as one of the most skillful and successful men in the profession. While in Colorado my father was haunted by an intuitive feeling that the gold-bearing quartz region of Alaska held a rich find in store for him. In October, 1882, a very strong corporation was organized in San Francisco, 'The Alaska Mining Co.,' to open and operate their extensive mines in Alaska. The directors of the company chose my father manager. They offered him an increased salary to go to Alaska to take entire charge of the work. This position he accepted and retained for five years. During that time he discovered a very rich mine on a small, rocky island near the coast. In partnership with his old friend, Mr. Dunbar, one of the San Francisco directors of the Alaska Mining Co., my father, at the end of five years service for the company, had developed the mine on the island into one of the best paying and most extensive of that famously rich gold bearing quartz region. This was the foundation and support of his vast fortune, which thereafter required his entire attention. At the death of Mr. Dunbar, which occurred in 1890, his one-third interest in the mine passed to his son, Dewitt C. Dunbar, a young man of great energy and integrity, with an excellent business education. He impressed my father as one in every way trustworthy and capable. At my father's request, Dewitt C. Dunbar, accompanied by his young wife, at once removed to Alaska. Under my father's tuition he began to prepare himself to take the active management of the mine, which had been christened 'The Martina.'

"In 1882, while on his first visit to San Francisco, my father met and loved Martina Morrison, my mother — my beautiful mother. She was twenty-seven, my father forty-two. They were perfectly adapted to each other, and both equally charmed and devoted. She possessed a fine mind, well cultured; a handsome physique, charmingly graceful in every movement; and, her crowning glory, an exceedingly amiable disposition. Martina Morrison, by those who knew her longest and best, was declared to be the soul of honor. She was an excellent medium, an enthusiastic and devoted Spiritualist—one of its purest and most eloquent exponents, highly esteemed by all as an able and earnest worker in the service of the two worlds. Fennimore Fenwick, my father, soon became much interested in her wonderful mediumship, and later became convinced of the absolute verity of the mighty truths of Spiritualism. He at once declared himself its willing and outspoken advocate: in his enthusiasm of delight he even hailed it as the coming religion of the world.

"Martina Morrison had such confidence in my father's future mining success, that she readily yielded to his urgent request for a speedy marriage, that she might accompany him on his first trip to Alaska. And thus it was they sailed away on their bridal tour, their destination that far off land of flashing glacier and unexplored forest, almost, if not quite, beyond the borders of civilization. This long voyage to an unknown country had no terrors for them. They were all the world to each other. A bright halo of hope and happiness spread a soft glow of enchantment over ship and sail, sea and sky, so vivid, so far reaching, that it even touched and tinted the distant shores of that far off, rock bound coast of Alaska. Smooth seas, lovely weather and favoring winds speeded the voyagers: those halcyon days flew swiftly by. Almost before they dreamed it possible the vessel came to anchor in the port that marked the end of the voyage. Safely landed, my father reported at once at the office of The Alaska Mining Company, only a few miles distant. There he commenced his five years of management for the Company, of which I have already spoken. There my mother remained until December, 1884, when she returned to San Francisco, to visit her friends. My father followed her five months later.

CHAPTER VII.

AN ALASKA KINDERGARTEN.

"IN JUNE, 1885, I was born, and soon became a very active member of the Fenwick family. I was pronounced by all who saw me an offspring in every way worthy of my noble father and my beautiful mother. When I was two months old, my parents returned to Alaska, taking me with them. There I remained until I was seven years old—seven years in that forbidding clime, so near the Arctic Circle. Isolated from other children, yet how happy and contented I was. Those years recall a troop of joyous memories, with not a bitter one to mar the group. My beloved parents were my only companions, playmates, teachers and confidants. I was papa's own girl. He was very proud of me and wished me to be with him as much as possible. He never wearied in the endless task of answering my questions, always so skillfully directing them by suggestions, that in my receptive mind there was soon unfolded a clear conception of the outlines of the different branches of all useful knowledge. When I was four years of age I knew the alphabet perfectly and could spell and construct a great number of words with my lettered blocks, and then copy them on my slate. When I was five years old, thanks to my mother's patient teaching, I could read fairly well. My father's ingenious methods soon made me familiar with the key-words of geology, chemistry, (including the names of minerals, metals and gases) botany, history, geography, physics and astronomy. I was unconsciously taught to associate these words or names with the groups, or families, to which they belong. I would spend hours with my father in the most delightful game of separating and classifying a miscellaneous heap of different colored blocks, bearing the names of minerals, metals and gases and the key-words of the studies I have just mentioned. To illustrate: The astronomy blocks were blue with the names in white letters; the geology blocks w re a deep reddish brown, with names in gray; chemistry, red, lettered in black; botany, green, lettered in yellow; geography, gray, lettered in blue; history, black, lettered in red; physics, a deep orange yellow, lettered in white; mathematics was represented in a small way by the cipher and nine digits, lettered in black upon ten plain unpainted blocks, giving in their forms that number of the principal geometrical figures, to which was added a shallow box with a broad lid, perforated by ten holes, corresponding to the blocks in number, size and shape, but large enough for the blocks to easily pass through into the box.

"In these groupings my childish interest and delight was intensified by my father's personification of the different families, such as: 'Mr. Astronomy Blue,' 'Mrs. Geology Brown,' 'Mr. Chemistry Red,' etc. For instance, the wonderful stories he told to me of the minerals, metals and gases—the sons and daughters of Mr. Chemistry Red, as he termed them—describing their loves and hates, the great variety of pranks they played, the queer combinations they entered into, the good and the bad work they performed, etc. These to me were fairy stories of the most charming kind, while at the same time they gave me a correct idea of the powers and properties of these unfamiliar things and served to identify them more closely as members of the chemistry family. My mother was a natural teacher, very proficient in botany, and in history, with its flower and fruitage of classic prose and inspiring poetry. She entered into my father's 'block-signal-system' of education with an enthusiasm as zealous and childish as my own, therefore her contributions to the rapidly increasing store of blocks were large and exceedingly interesting. Her stories regarding the numerous members of the botany and history families proved equally profitable and charming; those about plants and trees especially so. These stories and plays of science grouping, always associated with such pleasant emotions of my childish heart, became permanently fixed and dominant in my mental growth, forming separate brain structures around which the details of the accumulated knowledge of future years could easily and naturally classify and crystallize.

"Thus swiftly passed those happy years of my early girlhood. So constantly was I associated with my dear father and mother that schools I did not need. In my seventh year, under their supervision, I commenced a systematic course of scientific reading which I kept up until after I graduated from college. I commenced with the Science Primer Series, reading aloud to my parents one half hour each morning and evening, conversing and commenting on the different topics as we went along. This proved to be a continuation of the game of blocks: just as interesting, equally entertaining; all about the same familiar families I enjoyed it so much and never once dreamed I was accomplishing a great deal of good hard study. To me it was play; play that gave me more pleasure than any of my childish sports. I soon began to ask for an extension of the half hour lessons to an hour each; when my request was granted my cup of pleasure was full, my joy complete. With each succeeding week my interest in all my studies continued to grow. Yet my health remained perfect: my physical kept an even pace with my mental growth, largely owing, no doubt, to the much enjoyed hours of good romping exercise and the dancing and singing which followed my reading lessons.

"You must pardon me, Mr. Flagg, if I should tire you with such a detailed account of my child life; my excuse must be, the valuable hints it may offer when we come to consider a school system for the children of our model co-operative farm."

"I am profoundly interested," said Fillmore Flagg. "The very wonderful result flowing from the wise methods conceived by your parents and carried out by them so devotedly, fills my mind with admiration and offers a flood of suggestions as to the possibilities of what may be accomplished by a properly conducted, well equipped school on a co-operative farm. But you must not allow me to interrupt—please proceed with your very interesting story."

CHAPTER VIII.

AN INTERVIEW WITH THE "FAIRIES."

FERN FENWICK rose from her seat saying: "As it is near sunset, Mr. Flagg, I have something to show you in the way of a surprise, which I wish you to see before it becomes too dark: after having seen it you will better understand why this house was named 'Fairy Fern Cottage.' Therefore I propose that we now adjourn to the cool shade of the grounds at the rear of the cottage, postponing the recital of the remainder of my story until this evening."

"I shall be delighted to follow you," said Fillmore Flagg. "You have excited my curiosity; I am just in the mood to learn all I can about this lovely cottage and its beautiful surroundings."

As they reached the shady lawn, so cool and sweet

from its recent sprinkling, Fillmore Flagg observed that a wide, straight avenue, shaded by towering oaks and widely branching elms, led from the rear porch of the cottage to the broad front of the roomy stone stables, some two hundred and fifty feet distant. In the center of this avenue, with a finely graveled carriage drive on either side, rose a long line of huge stone arches, ten in number. These imposing structures of solid masonry were full thirty feet high, spreading to a width of thirty feet at the base. The two center arches were each twenty feet thick; the others, ten feet each. The open space between the arches was uniformly ten feet; the open circle under each arch was twenty feet in diameter. The vista formed by the spaces and arches together, was over two hundred feet in length. From the farther arch to the front of the stables lay thirty feet of smooth, clean gravel which covered, at this point, the full width of the avenue, seventy-five feet, forming the open court, around which was built the stables and the two tastefully designed stone buildings on either side—one, beautifully fitted up for the residence of the superintendent, the other containing the heating and pumping apparatus and the electric generator. The two wide center arches supported the huge metal tank which held the ample water supply of both cottage and outbuildings. Evidently, they were admirably adapted to that particular purpose. The rough stone work of the outside of all the arches was artistically covered and beautified by a luxuriant growth of intermingled ivy and cinnamon vine, which gave a still deeper shade to the interior. To the beholder, the exterior effect of the vines on the long line of arches was as beautifully romantic as if it really were one of those old Abbeys in picturesque ruin, so charmingly described by Sir Walter Scott. Deep grooves in the stone work, with light iron frames fastened near the outer edges of the arches, gave support during the cold weather to a roof of double glass, which covered all the open spaces between the arches, converting the whole into one vast greenhouse, through which passed the system of heating pipes from the furnace room to the cottage, thus providing a roomy winter home for an army of tropical plants and shrubs and at the same time protecting the water supply from the ill effects of all frost. A screen of interlacing vines, in place of the glass roof, now served to make the shade of the archway almost complete.

Having sufficiently examined the exterior and becoming to some extent familliar with the general plan and purpose of these unique arches, Fillmore Flagg and Fern Fenwick returned to the covered entrance from the kitchen porch. Here, as they were standing a few feet above the ground, they had an unobstructed view of the interior of the archway. Through the center, where the lower disc of the open circles touched the ground, ran a deep bed of coarse gravel, covered with a thick layer of smooth round pebbles, forming a perfectly drained pathway about three feet in width which extended uniformly from one end of the archway to the other. Conforming to the contour of the arches, rising and receding in unison, this pathway was bordered on either side by what appeared to be a continuous terrace of three stone benches, each one foot high and of the same width. These benches really were very heavy square terra cotta pipes, ingeniously cemented together with telescopic joints, and having thick, grooved covers which formed the protecting conduits for the wires of the lighting system and the pipes of the irrigating and heating apparatus.

Artistically arranged on these benches, in pots that were beautifully modeled, colored and glazed, was a wonderful collection of choice ferns, embracing all of the known varieties in prodigal profusion. The pots were so arranged that the smaller varieties occupied the lower benches, with the larger ones in gradually increasing sizes on the higher benches farther back. Viewed from either end of the archway they formed two matchless banks of the rarest verdure and the loveliest foliage the world ever saw. Everywhere the eye was delighted by great masses of drooping fronds of delicate green, like rare lace in fineness—outrivaling in beauty the plumes of the famous birds of paradise.

"This is simply superb!" exclaimed Fillmore Flagg. "I never saw anything one half so lovely! Shall we walk through now?"

"Wait a moment, Mr. Flagg," said Fern Fenwick. "The twilight shadows are so deep you have, as yet, caught only a glimpse of the rare beauty of my lovely ferns." Stepping quickly to the right side of the first arch, she pressed a button and lo! those wonderful banks of ferns, and all the space of the archway, was flooded with a glory of soft, clear light. A thousand tiny bulbs, in a lovely variety of flower and fern leaf patterns, gleamed and glowed from beneath the ferny banks or hung pendant, rainbow like, from the roof of this rock ribbed archway.

Held spellbound for some moments by his surprise, admiration and delight, Fillmore Flagg murmured softly, almost in a whisper: "Can anything surpass this vision of perfect beauty?"

"Yes," said Fern Fenwick, radiant and smiling,

"I think it can be surpassed, but we must allow the enchantress to use her magic once more, by giving my darling ferns their bath of beauty. Then you shall see them in their diamond robes."

Saying this, she pressed another button. A thousand tiny pipes, concealed in the ribs of the stone roof, gave forth a shower of fine spray, filling the long fernery with a hazy mist of cobweb fineness. Very soon millions of globules of moisture gathered on leaf, stock, frond, plume and tiny tip of every leaflet, reflecting each ray of light with diamond-like brilliancy. Pressing another button to shut off the spray, Fern Fenwick said:

"Now, Mr. Flagg, my ferns have donned their royal robes and are ready for your tour of admiring inspection. I assure you they are worthy of it. As a choice collection of ferns in such perfect condition, its equal cannot be found in all the wide world! As a collector I am an enthusiast; for many months I have travelled far and wide in my efforts to add new specimens of rare beauty to the original collection. You may guess how much I prize it when I tell you that money could not buy it."

"You are surely a most wonderful enchantress," replied Fillmore Flagg. "I feel that under the potent spell of your magical wand, I have entered the inner mysteries of some glorious temple of ferns, in a world of enchantment! I am so fascinated and dazzled by this marvellous display of brilliancy and beauty, that I am moved to pay homage to you, Miss Fenwick, as a fitting tribute of loyal devotion to Fern, the Fairy Queen of this fair temple."

As he finished his gallant speech, the deep tones of emotion vibrating in the full rich voice of Fillmore Flagg, and the look of intense admiration which shone so eloquently from his eyes, brought a flush of color to the fair face of Fern Fenwick and warned her that it was time to be moving. Skillfully keeping up the personification, she quickly said:

"Mr. Flagg, I am delighted on behalf of the fairies to express thanks for the glowing tribute to their Queen which you have so beautifully voiced. Let us now walk through to the end of the fernery and return. As we pass along I will point out my favorite plants."

Only a few steps had been taken when Fillmore Flagg paused, listening and looking about him in all directions, with a very puzzled expression. A delightfully cool breeze was fanning their faces: this breeze was laden with some strangely sweet perfume both soothing and stimulating to the senses. The air all about them seemed to vibrate with the distant melody of some angelic music, now sinking, now swelling in perfect harmony; so soft, so clear, so bright, so inspiring in its wealth of tone and joyous movement.

"Ah! Miss Fenwick," said Fillmore Flagg, "my senses are all entranced! Your wonderful fairies in this grotto of magic are at this moment thrilling my being with sensations of the most intense delight! How can the Fairy Queen explain? What has she been doing with her magical wand to produce such delicious perfume; such entrancing music?"

Fern's merry laugh rang out musically clear, and her eyes sparkled roguishly as she replied: "I assure you Mr. Flagg, that in this instance the fairies are not responsible. The explanation is quite simple but rather long. Therefore let us move forward while I give you the details: As we were stepping down on this graveled walk, I turned the switch and started the ventilating fans, at the same time connecting the electric current with a series of melophones located near the top of the arches. Along the ventilating tubes, in a series of small compartments, are sponges saturated with different kinds of perfume. These sponges can be exposed to the air current or withdrawn at will, yielding a single perfume or a blending of as many kinds as one may wish. The wonderful variety of these choice blendings, which can be so easily produced, affords a constant succession of sweet surprises. The melophones which you hear, represent the highest achievement of art in the production of automatic musical instruments. This set is the most complete and the most expensive one in existence. In construction and final completion they cost the inventor and maker three years of constant thought and labor. The result is truly marvellous. The perfection of harmony and purity of tone are convincing testimonials of their excellence. In operation these instruments are placed in a very large double tube made from a peculiar kind of metallic alloy recently discovered, which affords the most perfect conditions for the conservation and conductivity of all musical vibrations. They are capable of producing an almost endless variety of choice music. The selection which we hear at this time, is one which I have re-named 'The Carol of the Ferns.' Pardon me, Mr. Flagg, if in my enthusiasm over the beauties of what you have so poetically termed my 'magical temple of ferns,' some of my statements should sound like boasting; I assure you they are not so intended. I trust that now I have cleared up the mystery to your perfect satisfaction."

"Charmingly," said Fillmore Flagg, "Nevertheless my fairyland illusions still abide with me; I confess

I am still under the spell of the great happiness they have given to me—I shall never forget it. The truth in this case proves even stranger than fiction; I quite agree with you that in all the wide world there is nothing like this! It seems to me that those extraordinary melophones yield the finest music I have ever heard. In sweetness and purity of tone, softness and wealth of harmony, which is pervaded by some electric quality of inspiration, so stirring, so thrilling that every nerve and every cell in the body responds. They stand unrivaled as the very acme of musical art. I now understand why your lovely home here should be named 'Fairy Fern Cottage.' I fully appreciate the significance of the title. This royal temple of ferns makes the name most fittingly appropriate, and easily ranks this cottage as the eighth wonder of the world! The fame of its rare beauty should be known in every land. You ought to be very proud of it. I assure you, Miss Fenwick, that you are abundantly justified in praising it enthusiastically at all times, without fear of being considered egotistical. But tell me, if I may be permitted to ask, who was the wonderful genius who first conceived and planned the building of this imposing line of arches? So useful, so ornamental, so unique, yet so perfectly adapted as a summer and a winter home for your ferns and flowers and, withal, offering such a perfect title to your unrivaled cottage home."

"Thank you, Mr. Flagg, for that question. In my reply I am eager to pay a deserved tribute to the dearest and noblest of men—my father. Inspired by his love for me, his brilliant mind conceived the entire plan and purpose of this curiously novel structure. He succeeded in completing it and also in filling it with the original collection of ferns, without my knowledge. On the morning of my fifteenth birthday, he brought me here to bestow upon me this priceless gift. The surprise was a perfect one. When he made me understand that he gave with it a deed to the cottage and grounds, the surprise became so intense that it fairly took my breath away. I was so overjoyed that by turns I laughed, and cried, and hugged papa, until I came very near to having a genuine fit of hysteria! At that time we changed the name of the house to Fairy Fern Cottage. This is why I am so proud and so fond of my cottage home. This is why I appreciate your praise of it so much—why I am so thankful for it. I feel sure that you will now appreciate my sincerity when I repeat that money could not buy it!"

CHAPTER IX.

THE PROBLEM VS. A GOOD MAN WHO IS AS RICH AS HE IS NOBLE.

AFTER supper Fern Fenwick and Fillmore Flagg returned to the tower room for the continuation of the story. She began by saying:

"Let us return to my father's mining operations in Alaska. In 1892, Dewitt C. Dunbar assumed the active management of the Martina mine. A large proportion of my father's surplus capital from the mine had been invested, through trusty agents, in the cities of San Francisco, Saint Paul, Chicago, Washington and New York. We at once planned a tour of travel that would give him the opportunity to personally inspect these investments, and at the same time give me a chance to see the world, and to mingle in society, or so much of it as a continuous hotel life might offer.

For my mother and myself this delightful tour was one long holiday. We enjoyed it so much. To me especially, it proved exceedingly profitable; geographically speaking, my ideas of the largeness of the world, and the vast number of its people, were wonderfully expanded. In December, 1893, father completed his investments by the purchase of a winter home in the city of Washington, and this summer home here. This cottage was built in the year 1900.

During the summer of 1894 we visited the brothers and sisters of my father, who were at that time living with their families on farms in Iowa, Minnesota, Nebraska, Kansas and Missouri. As was generally the rule, with a large class of farmers in those states at that time, we found them, with but few exceptions, poor, in debt, and very much discouraged by the menacing outlook for the future. Farm interests everywhere were in a desperate condition. A succession of twenty years of falling prices for all farm products, accompanied by frequent calamities, such as hail storms, hurricanes, hot, blighting winds, drouth and armies of grasshoppers, had so multiplied and magnified the farm debts, and so reduced the value of farm, stock, and product, that even the interest on the indebtedness could no longer be kept up; ruin and beggary threatened the entire community of farmers. Under the severe pressure of these conditions, great numbers of the more unfortunate abandoned their farms in despair and sought employment elsewhere, mostly in manufacturing centres and the large eastern cities. Much of the money and wealth of the land had flown to those points, thither logically, they followed, to enter the ranks of that vast army of competitors for the crumbs that might fall from the table of an already glutted labor mart; to learn by bitter experience now cruelly the system of competition in all kinds of business can grind the helpless poor; to learn, through years of suffering, the real meaning of competition, that so long as it rules over commercial and industrial systems, the rich must grow richer and fewer in number, while the poor must grow poorer, and more and more numerous; to apprehend, slowly and painfully, that by coming from farm to city they had still farther congested the already overstocked labor market, thereby adding fierceness to the competition, insuring an increase in the purchasing power of the dollars of those who held the labor market, while they correspondingly decreased the possibilities for earning the dollars they must have in order to live; to perceive dimly in their desperation, that congestion of the labor market speedily affected all markets; that an overstocked labor market always meant a decrease of wages, which in turn, caused a corresponding shrinkage in the number of purchasers for all salable goods in the general market, followed by increased panic and stringency in the money market; which speedily rolled up another disaster, sweeping in turn, additional thousands into the ranks of the unemployed; demonstrating, finally, that a repetition of these evils is inevitable; that competition in its last analysis, means the complete destruction of all business.

"As my father came to understand the full significance of this deplorable situation, involving and distressing his own brothers and sisters, his noble nature was grieved and shocked. He made haste to place his people in a condition of financial independence. How happy and grateful they were! And my father rejoiced with us that he was able to offer such timely assistance. He then announced to us his determination to devote the remainder of his life, and so much of his fortune as might be necessary, to the solution of the problem of how best to overcome the blighting evils of the competitive system. After much thought, long research and hard study, he decided to commence with the land as the necessary basis of all progress; with the farm as the rational progressive unit; with improved farm methods on co-operative lines, as the lever by which to restore the control of the land to the farmers, and to lift them and their sons and daughters from the class of ignorant dependents, to a class of cultured independents, which should be well worthy of serving as a model in the race of progress, for all the other classes. In his efforts to modify, correct, and reform social and business methods, he proposed to use the strong and kindly arms of Co-operation in fighting the evils of Competition, or its representative, the pitiless competitive system. He reasoned that all forms of government are but the result of co-operative effort. Both experience and observation had taught him that the measure of excellence of any government is the measure of its perfection in co-operation. Therefore it logically follows, that the more perfect the co-operation achieved by the administration of any form of government, the greater the degree of justice and equality attained in the distribution of benefits to all of the governed."

CHAPTER X.

THE REAPING OF THE DEATH ANGEL.

"TOWARDS the close of the summer of 1895, my father placed me in the preparatory department of Vassar College, where I made rapid progress. I began to appreciate the superior wisdom of the methods of teaching which my parents had so systematically carried out for my improvement. Thanks to their efforts, I held the key to all of the sciences, history and literature, prose and poetry! All of their principal words or terms with their definitions, were familiar friends to me; while all new facts regarding their various subdivisions, auxiliaries, etc., and the relations existing between them as such, were matters of absorbing interest to me; so much so, that I soon became master of the subject I was studying, very often proving a puzzling surprise to my teachers. At the age of twelve I entered the regular course and graduated from college just as I was entering my eighteenth year, being by four years the youngest member of a graduating class of one hundred girls.

"Some months after my fourteenth birthday, my darling mother was taken from me in the mortal form, very suddenly and most unexpectedly. My father was away from home on a long trip to Alaska. I was at Vassar. My mother was with a congenial party of friends at a favorite seaside resort. One day while bathing, one lady of the party swam too far out, was taken with a cramp and shrieked for help. My mother, who was nearest, being an excellent swimmer, courageously went to her assistance. Unfortunately, the tide was running full and strong and was against my mother in her heroic struggle to save her friend. Alas! before aid could reach them both sank beneath the waves and were lost. My noble mother had generously sacrificed her earthly existence in her brave effort to save the life of another! This was my first experience of the grief and desolation that follows the reaping of the Death Angel. In my youth, my half-dazed condition, I could neither realize nor understand what later became so plain to me; that to die is to live again. That death, so-called, is but the change from one form of life to another, which is still higher in the scale of progress. Nor could I then realize, that for the purpose of bringing to me a consciousness of the possibilities of my spiritual being; under the ministrations of the angel of compensation, out of the very depths of the gulf of bereavement and sadness through which I was passing, there was coming to me the precious gift of a priceless mediumship, the marvelous key! the all-potent 'open sesame' with which to unlock the gates between the two worlds and reunite the separated loved ones on either side.

"At that time Mrs. Bainbridge, then but recently widowed, was in charge of the old home here. She was an excellent medium who had often proved herself worthy of my mother's entire confidence. Acting under the guidance of my arisen mother, she at once, without hesitation, took charge of all business arrangements, especially those of preparing for the cremation of my mother's body, in accordance with her often expressed wish. She telegraphed the sad news to my father in Alaska, asking for instructions. He replied at once that the body must be cremated, as my mother had directed in her will. He would return as soon as possible, but at the best he could not hope to arrive in less than two months. In the meantime, Mrs. Bainbridge was authorized to take entire charge of 'Fern,' and of his business affairs that needed attention, until he came.

"I came home from college, sorely grieved and shocked at the awful suddenness of my mother's transition, but through the mediumship of Mrs. Bainbridge, my mother, having her in a deep trance, was soon able to comfort me; to make me realize that she was not dead, but still near me with all a mother's love and tender care. From time to time she directed Mrs. Bainbridge how to manage the pressing business that came up. She told me that she had long known that I was endowed with wonderful mediumistic power, which must now be fully developed for her sake, as a necessary and natural channel of communication so desirable to her, which she should prize very highly. Also as a source of comfort for myself and my father, especially as a joyful surprise for him when he came home. Therefore it was decided between us that I was to sit one hour each day with Mrs. Bainbridge for development. My mother seemed to feel sure that I would make an excellent trumpet medium, and encouraged me by predicting my speedy development as such. Strangely enough, so it proved. My progress was rapid. In two weeks time my mother could speak to me through the trumpet without difficulty and much to my delight. I began to appreciate the great value of my wonderful gift and to understand what it meant. Our dear family circle, which in my despair I had thought broken forever, was now reunited. Father, mother, daughter! just us three as of yore. And—the wonder of it—I, the youngest, the weakest and the least wise of the trio, was the instrument!

When I thought of the possibilities, of the joy and consolation it would bring to my father and mother, my heart swelled with gratitude and thankfulness that this mighty power had come to me. The power to destroy the dread of death; to demonstrate the continuity of life; to prove that the binding love of family ties, kindred, and cherished friends still shone with untarnished lustre beyond the shadows of the silent grave. How beautiful, how wonderful, how glorious it was! And with this power came the solemn charge that I was to cherish it with care and keep it pure and holy. Yes, I resolved that I would do this conscientiously. It should be my highest ambition to ever use my mediumship with my best and most unselfish aspirations, to keep it apart from the grosser things of life, to dedicate it to good and to good alone. And thus it was that my mediumship continued to develop and grow in perfection. My mother could talk with me as often as she wished and as long at each sitting as she desired. I was no longer alone or despondent, my darling mother still could be, and was really, my mentor, friend, parent, teacher and spiritual guide. I forgot to mourn or to feel lonely, though I longed for my father's home-coming that we might share this new found joy. So interested was I and so occupied, that the two months quickly passed and my dear father reached his home in safety. I had arranged for a quiet evening with him alone. When my mother, through the trumpet, joined in the conversation and welcomed him with loving words of endearment, so familiar in the greetings of other days, he was almost overcome by the flood of ecstatic emotions that moved and thrilled him as he began to appreciate the significance of such a miraculous surprise. His heart was glowing and his entire being permeated with this great wave of happiness. His face was radiant with joy and beamed with fatherly affection and pride as he pressed me to his heart again and again, thanking me for my thoughtful spiritual work in the development of my wonderful gift, which, for his consolation, I had striven so unselfishly, so ardently and so earnestly to attain, while facing alone the one great crisis of my young life. Still holding me in his arms, he looked into my eyes long and fondly, almost adoringly, as he said: 'With such a daughter, whose loving heart and purity of soul has won for her the marvellous power to reunite our broken family circle, I am indeed the most fortunate of all men.' Then in a moment I perceived that I was no longer a child, I was a woman; that henceforth my father would think of me as a woman—still his loving daughter—but also his equal, his confidant, his trusted friend, his adviser in times of need, his oracle, his medium of communication with the loved ones who dwelt in the world of spirit. How good and beautiful was life in the light of this new vista of possibilities and responsibilities for me! For the moment I seemed to be transported to some grand spiritual height, where as a responsive spiritual unit, I felt the throbbing of the limitless sea of environmental life surrounding me like a golden mist, on every hand. Every pulsation proclaimed my immortality as a part of that boundless sea; boundless, fathomless, unthinkably shoreless! of life, all-producing, all-containing! My soul no longer questioned. It was filled with a peace and joy that passeth the power of words to describe.

"Thus inspired and encouraged for the future, I was ready and eager to take up again the active duties of life. In resuming my collegiate studies, it was agreed between my father and mother and myself, that I should come home from Vassar every Friday evening, returning by the early train Monday morning, the intervening time to be sacredly devoted to our trumpet family circles. Oh, Mr. Flagg! How happy we were then! For the next three years nothing was allowed to interfere with these delightful reunions, whose memories are associated with so many incidents that bound us three so closely with the silver cords of pure affection.

"After leaving college, I accompanied my father in all of his journeyings after new data in economics and agriculture. For this purpose we spent the winter of 1902–3, travelling in France, Italy, Germany and England, returning to America in April, 1903.

CHAPTER XI.

THE MARTINA MINE.

"Early in June of the same year, Dewitt C. Dunbar discovered a new lead in the Martina mine which proved to be of such marvelous size and richness, that my father's personal inspection was demanded at the earliest possible moment, to decide on the best methods of pushing forward the new work, and also to determine what part of the old work should be continued. The numerous letters and telegrams from Mr. Dunbar, all urging the utmost haste on my father's part, gave him but little time to consider the results of such a long journey, or to make the proper preparations for it. It was evident that Mr. Dunbar must be in a state of intense excitement. In order to catch the next steamer from San Francisco, father left a number of important items of business for me to transact. I wished very much to go with him but all the circumstances seemed to conspire against me. Father promised to return at the earliest possible moment, meanwhile he was to send me a dispatch announcing his safe arrival in Alaska. By the end of July, messages, and later, letters began to reach me announcing the wonderful output of gold from the new lead. So rich was the ore that for a time it was thought best to abandon all work in the old mine. I could see very plainly from his letters that the fever of Mr. Dunbar's excitement and enthusiasm had also claimed my father as a victim. I then foresaw that his stay in Alaska would be prolonged far beyond my expectations or his own. I began to feel very uneasy and to wish most fervently that I had insisted on going with him. I resolved in future to keep him company wherever he journeyed. Meanwhile the yield of gold from the new lead continued to increase. The value of the Martina rose like magic; offers to purchase at fabulous prices came pouring in. Mr. Dunbar would not accept, and decided, then and there, to remain another ten years as manager and resident superintendent of the mine. That settled the question. After that, my father announced that the mine was not for sale at any price. In writing to me concerning the matter, he says:

"'My Dear Fern: * * * I at that time decided that my interest in the mine which I had named for your mother, and which had proven the luckiest and richest in Alaska, should pass to you as it came to me, entirely unincumbered. So rest assured, my daughter, so long as Dewitt C. Dunbar is able and willing to manage the mine, both my interests and yours are in safe hands; in skill, honesty and ability he is one of the grandest men I have ever known; he is a treasure. You can trust hlm implicitly!'

"As I had anticipated, it was December before my father could leave Alaska. In a letter dated Dec. 5, to which I shall again refer, he says:

"'I have planned to leave here on a steamer that sails on the tenth of this month. I fear the voyage may prove a rough one. I have a foolish dread of it, which is quite unusual for me. I am oppressed by an uneasy feeling which I strive in vain to shake off. However, I have taken good care to make such arrangements with Mr. Dunbar as will cover all possible contingencies. This is to be my last trip.'

"On the twelfth of December I received a message from Mr. Dunbar, stating that Fennimore Fenwick had sailed on the tenth as he had planned; that he was well and strong, and would wire me as soon as he reached San Francisco. This cheering message gave me new courage, I began to count the days and to look forward more hopefully. I decided, although it was so late in the season, to wait here in the cottage until my father came. When Mrs. Bainbridge left to open our house in Washington, I had intended to follow her a few days before Christmas, but for some unexplained reason, I could not make up my mind to leave the cottage. After the message came the question was settled—I was to remain here.

CHAPTER XII.

SPIRIT AND MORTAL.—FATHER AND DAUGHTER.

"AT THIS point, Mr. Flagg, I wish you to carefully note the significance of the strange event which soon followed. Christmas Eve, 1903, found me here alone, seated at my desk, alternately reading, musing and writing. All day a terriffic snow storm had been raging, at nightfall it continued with increased severity. I could hear the fierce gale shriek as it lashed the tree tops furiously. I shuddered when I thought what danger such a gale might mean to the good steamer, bearing my father homeward bound across the rough, icy waters of that far off wintry sea; that yawning, terrible, treacherous sea!

"During the afternoon I had been nervous and lonely. As a solace, I had a long talk from my mother through the trumpet, which cheered and comforted me greatly, especially her confident promise that I should hear from papa even sooner than I had hoped. Over this I was musing when a strange thing happened. I was startled by the low tones of a familiar voice from the trumpet. Almost frozen with fear, I heard: 'Do not be frightened, my darling; I am your father, Fennimore Fenwick, who loves you, if possible, more than ever. A frightful storm wrecked the steamer and released me from my body. Nearly all of the passengers and crew perished with me. A few still survive; they are in a single open boat, tossing helplessly in the awful surge of that wild waste of water, possibly they may yet be saved. My dear wife, Martina, your own beautiful mother, was watching and waiting for me at the scene of the wreck. Hers the beautiful arms that welcomed me as I was born into the new life of the spirit. How glorious it was that she, so dear to me, could be there. In the radiance and splendor of all her spiritual loveliness, I was charmed almost to the point of forgetfulness. I seemed to be floating on the bosom of a sea of golden mist, my spirit filled with a measureless contentment. Presently I awoke to a vivid consciousness of my new life. In the light of the loving eyes of my peerless Martina, I was soon made to realize that I had just passed painlessly from life mortal to life spiritual. I perceived that time and space no longer barred the flight of my freed spirit. Hand in hand we came; almost before I knew it we were here. Thanks to your mediumship, and to this trumpet, I could come and speak to you so soon. Yes, my dear child, we three, a loving trio, are still united just as of yore. I shall be permitted to help you, from this side of life, to carry out and complete my plans and purposes regarding improved modes of farm life. I wrote you from Alaska on the fifth of this month, announcing my intention of sailing on the tenth that letter came by a Victoria steamer and will soon reach you. At that time I was weighed down by a premonition of some impending disaster. So seriously was I impressed, that I at once made arrangements with Dewitt C. Dunbar, in case of my death, to continue to operate the mine in partnership with you on the terms now in force, and this he was perfectly willing to do. By the terms of my will, now in the hands of my attorneys at Washington, you are at this moment, sole heir to my large fortune. As you know, I long ago placed my brothers and sisters beyond the reach of want. Well do I know, my dear girl, that I can trust you perfectly, to carry forward my work.'

"As his voice ceased to vibrate in the trumpet, I sprang to my feet with outstretched and imploring hands: 'Father!' I cried, 'How can I do this work alone? I am yet but a child, with a very limited business experience to fit me for this great responsibility.' He at once replied: 'Fear not, my child. Faithful, capable, and trustworthy help shall be brought to you. At all times I shall be near, to advise, and to guard you and your interests. Go forward bravely in the conscious power of your own potential spirit, dominant and dauntless. Armed with the majesty and mystery of your mediumship, all obstacles shall yield, and naught shall prevail over you!' This prophetic command, so thrilling, so imperative, touched and stirred my inner self; my soul responded to the appeal. In one brief moment I regained my self control; was calm, could think clearly and reason logically.

"At intervals throughout the night I continued to consult with my parents. My father advised me to write at once, announcing his death, and requesting Mr. Dunbar to fix a time at which he could meet me in San Francisco, for a conference. This I did at the earliest practicable moment."

CHAPTER XIII.

QUESTIONS AND ANSWERS.

AT THIS point in her story, Fern Fenwick said: "Mr. Flagg, I now realize the wonderful prescience of my father's promise of abundant and timely help, especially when I consider your life work, and the masterly way you have equipped yourself for it, and finally, by the mysterious manner in which we were brought together. Is it not almost like a miracle?"

"Really, Miss Fenwick, I am lost in amazement! It seems to me that I must be dreaming! The situation is so entirely outside of my experience, so unthinkably strange to me, that I doubt my ability to discuss it intelligently. Your story is the most marvelous of anything I have ever heard. I feel quite sure that it must be strictly true, yet I can scarcely comprehend it. A host of questions arise in my mind, which I wish to ask, if I may be permitted. When you heard the voice from the trumpet, how could you feel so sure it was your father speaking? That he had been swallowed up by the sea? That the shipwreck had really occurred?"

"I do not wonder at your questions, Mr. Flagg," said Fern Fenwick, "Iwill gladly answer as best I can. Without considering or discussing the fact that the crucial test of identity was disclosed by almost every word which my father uttered, yet I could not for a moment doubt his presence. I knew he was there. I recognized every intonation of the voice. I felt the identity of his spiritual personality, radiant with the silent force of his love for me, quite as plainly as though at that moment his physical personality had entered the room. My experience after my mother's transition, the development of my mediumship, and my increased sensitiveness to the presence of spiritual entities, no doubt aided me greatly. At that time I perceived and recognized without question, that life in the physical is but the expression of the spirit, or Ego; that after the passing of the physical, the Ego inherits and possesses immortality as a conscious individual entity, clothed with a spiritual body, perfectly fitted for its continued existence in the realms of the world of spirit; that, through the action of a natural law, the law of mediumship, such spirits can and do, come to and communicate with their friends and loved ones in earth life. All these things, I knew my father understood clearly, therefore I was prepared to accept the verity of his spiritual presence as readily as I would any other phenomenon of nature. In conclusion, I may as well tell you at this point, that the letter referred to by father as having been written by him in Alaska on December fifth, together with my conference in San Francisco, some months later, with Dewitt C. Dunbar; the arrival in port at that time of a China steamer, bringing the mate and four sailors as sole survivors from the wreck of the ill-fated steamer, and my interview with them, all confirmed, in every particular, the truth of the statements concerning the matter, which were made by my spirit father, just after his passage through the gateway of death from life mortal to life spiritual. Can I add anything more convincing?"

"Pardon me, Miss Fenwick! I believe what you have told me is absolutely true. I can perceive and appreciate its wonderful significance only in part. I understand now clearly why it was necessary for me to know so much of the story of your life and that of your noble father. I have listened to your story with almost breathless interest, with all I am profoundly impressed. A new world is opening to me. My mental and spiritual horizon has been extended beyond the power of words to express. Life has a thousand new meanings: In them I read the importance and responsibility of the great work we are about to undertake. I wait with increased interest for my personal interview with your father. Now that I have heard so much of him, I bow with added reverence to his great and noble love for humanity which prompted, and his wonderful genius which conceived and planned the work so generously. I am proud and thankful that I have been chosen as an instrument deemed capable and worthy of helping to carry it forward.

"As to things spiritual, pertaining to a life beyond the grave, I am intensely interested and eager to know more. May I hope, Miss Fenwick, that you will kindly consent to become my teacher in this new school of wonderful phenomena and spiritual law? I too, am alone in the world; my father and mother have both passed the bitter flood of the dark river of death. They too, like your parents, must now be living in the world of spirit as conscious, loving father and mother, with hearts filled with a living, glowing affection that can and will respond to my own. Can it be possible that I am to feel and know this by direct communication with them?"

"I shall be delighted, Mr. Flagg, to help you in this matter in any way that I can. Your desire for a direct communication from your parents is perfectly natural and right and, I doubt not, will be fully gratified in a few days.

"In this connection, let me ask: Have you ever had a seance with a medium? Do you know anything about the laws that control and govern mediumship? Have you been interested to any extent in reading the all-comprehensive philosophy which mediumship demonstrates?"

"I am very glad, Miss Fenwick, that you have put those questions. I desire to state briefly and frankly my attitude, up to this time, towards mediumship and the philosophy and phenomena of spiritual manifestations generally: I believe I was a born agnostic. All my life I have been skeptical as to the verity of a life beyond the grave. In this I have differed widely from my people, a large majority of whom have been zealous Presbyterians for at least five generations, while I have followed Voltaire and Ingersoll. In the ranks of their following I have been content to cry: 'I don't know! I can wait! One world at a time is enough for me!' As to mediumship, or any manifestations of it, I know almost nothing. The few mediums I have met accidentally, have unfortunately failed to impress me favorably. All that I have heard or read of them has had a strong tendency to predjudice me against them and the philosophy they taught. Therefore, until my visit to this cottage, I have never been at all interested in the matter. I now perceive that in studying the great problem of life, and how best to learn most about it, I have utterly ignored one of the most important sources of both information and inspiration. My predjudice and indifference have vanished. I wonder at myself, at my readiness to acept your point of view regarding your most marvelous mediumship and its wonderful manifestations; at my feverish interest and anxiety to learn all I can about things spiritual at the earliest possible moment; at my intense longing for the complete verification of all the beautiful propositions relating to spiritual life which you have stated so eloquently and so convincingly; but most of all do I wonder and am amazed that these things are not miracles; that they occur through the action of natural law, which, if true, makes it possible—nay probable—that mediumship and its manifestations are as old as life itself. This,

Miss Fenwick, defines my position as clearly as I can state it. Do you think I am likely to prove a pupil worthy of his teacher?"

"I most assuredly do, Mr. Flagg," said Fern. "I think you are now prepared for the promised interview with my father. However, before he joins us, I wish to say by way of explanation, that when I am here alone, he can use the trumpet with ease at any moment and in any kind of light, but in the presence of strangers, different conditions are required. We shall at first be obliged to use another kind of light. By the aid of this light you can plainly see the trumpet, supported horizontally in the air just over his chair, but you will be unable to discern even the faintest outline of the spiritual form holding it; as in using the trumpet, the vital force of both the manifesting spirit and the medium is concentrated in the trumpet in the effort of speaking. Sit perfectly quiet for a moment; I will close the windows and prepare the room."

A few touches on the small keyboard in her desk, and lo the heavy double curtains swiftly and silently unrolled and covered the windows. At the same moment, the beautifully ornamented, dome shaped center of the lofty ceiling began to glow with a constellation of soft, phosphorescent lights, filling the room with a radiance as mild and silvery as moonlight, and yet even more soothing to the nerves. Presently the air was vibrant with the low, sweet strains of distant music, soft and slow and of such exquisite harmony that it seemed a rare combination of all that was inspiring, charming and beautiful in the variations of time, sound and rythm. The combined effect of the light and the music on Fillmore Flagg was electrical. Every nerve was thrilled with rapture. He was completely absorbed. As the music ceased he turned with a start to look for the trumpet. As he looked, it slowly rose from the chair and there came from it the clear tones of a manly voice, full of sweetness and power. He heard these words: "Fern, my daughter, will you tell this gentleman who I am?"

"My dear father," said Fern, "How glad I am that you have joined us! Mr. Flagg, this is my father, Fennimore Fenwick, of whom I have told you so much. Father, this is Mr. Fillmore Flagg, who, as you already know, has promised to devote himself to our work."

As the trumpet slowly moved nearer, Mr. Fenwick said: "Mr. Flagg, as the father of Fern Fenwick, I extend to you a cordial greeting and a most hearty welcome to Fairy Fern Cottage. I trust this is but the commencement of a long and uninterrupted acquaintence, which may soon ripen into a true friendship, that shall bring much pleasure and profit to both. I am exceedingly well pleased with your advanced ideas on the subject of co-operative farming as the proper cure for the evils that now make farm life so miserable and so unsatisfactory. I wish particularly to congratulate you on the thoroughly systematic and successful methods you have adopted to fit yourself so well for this peculiar work.

"Now my young friend, one moment to another matter which is likely to prove of great interest to you. I find your parents in spirit life. I met them since you came to the cottage. They approve of your chosen life work. They are very proud of you, their beloved son and only child. They bid me give you a message of love with the assurance that they will speak to you through this trumpet very soon."

"Mr. Fenwick," said Fillmore Flagg, "I thank you for the encouragement of your kindly greeting and for the many pleasant things you have said of me and my work. In the future I shall strive conscientiously to merit your praise, and hope to earn your lasting friendship. As to the glad tidings from my parents in spirit life, I am rejoiced. In my heart the torch of hope is lighted; its pure flame is fast burning away the barriers of the belief I have so long entertained, that 'Death ends all,' also of the equally depressing creed of my Presbyterian people, who have so long taught and thought that 'The dead know not anything;' that my parents, with that vast army of souls, having passed the portals of the tomb, are now lost in the oblivion of that long unconscious, dreamless slumber, which stretches from the new made grave to The Day of Judgment. Hence, the message of love from my parents, with the assurance that they will speak to me so soon, has made me very happy. I am content to wait patiently for such further messages as opportunity may bring to me. I am ready and eager, Mr. Fenwick, to hear your plans. Please proceed."

"Very well," said Fennimore Fenwick. "Fern, my daughter, you are to remain at your desk with pencil and note book, prepared to take down what I have to say."

CHAPTER XIV.

THE ETHICS OF PLANETARY EVOLUTION.

"In order to plan this work wisely, and to discuss it understandingly, it will be necessary at the beginning to go back to first principles, to try to discover the real object and purpose of human life on this planet. In searching along the pathway of countless ages in our planet's history, we discover a continuous upward movement in the progression of the manifestations of life; from the mineral to the vegetable; from the vegetable to the animal; from the animal to man. Man representing the apex of progress in the constantly ascending spiral of the evolution of life from the birth of the planet to the present time. Therefore, both spirit and mortal, we are all children of the planet, chained to its destiny, all alike working factors in the achievement of its purpose so mighty. Through the planet, its solar system, and the system of systems in a long line of an infinite series, far beyond the power of computation, we are also the children of the Great Oversoul, the Source and Center of all life!

"Human life, then, is the flower and fruit of the planet—the highest combined expression of its life—each life a planetary seed, a concentrated possibility of all expressions of planet life. Perhaps the most convincing and beautiful illustration of the truth of this vital and all important proposition is, that the reproductive cells of man in his highest state of development, multiply by fission, or self-division into halves, as did the primal sperm of protoplasm at the very beginning of vegetable and animal life. This great philogenetic vine with its myriads of branching arms, reaches in an unbroken line from the lowest to the highest forms of life; all alike are fruit of this vine. This offers indisputable evidence of the common brotherhood of humanity! the motherhood of the planet! the fatherhood of the Great Oversoul!

"From these premises we may safely conclude that the object and purpose of this planet is the evolution of human beings, their continued growth and development, until the state of perfection for the entire race is reached. With this comes the complete achievement of the purpose of the existence of the planet. Hence, we perceive that human life is the most precious production of the planet. Henceforth its energies are to flow towards the perfecting of the human race.

"In the great, white light of a higher understanding of these basic and vital truths, let us strive to make conditions for the protection of ALL human life. The task becomes less difficult as we more readily comprehend and appreciate the magnitude of the thought, that through the planet, this sacred life is the immortal and enduring expression of the Eternal Spirit. Viewed in this light, we apprehend clearly that all acts, by society or individuals, which tend to protect, promote and purify this life, are good, right and holy, and in their doing, become the highest and best expression of a sacred religious duty. On the contrary, all acts of society or individuals, which tend to destroy, injure, poison or sully this sacred life, or to bar its ordained progress, are, in themselves, unholy, wrong, criminal and cruel, and in commission, become the greatest and most unpardonable of all sins.

"All this becomes more apparent, when we consider that the sum of the pleasant sensations of the individual, and the happifying emotions which flow from them, constitutes the sum of human happiness. All conditions of life which promote right living, ethical culture and moral growth, nourish and call forth emotions of truth and honesty, pure pleasure, adoration, worship, hope, affection, love and all the higher and nobler characteristics, build up life and increase its capacity for happiness. Through the action of an equally inexorable and unswerving law, the misery and crime which poverty breeds, with its bitterness of hate, grief and despair, and all the train of other evil emotions engendered thereby, are poisonous in their nature; they tear down and destroy life. Therefore that social and industrial system which affords most abundantly, and for all of the people, conditions that are life-promoting and poverty-banishing, is logically the nearest just and right, because it is the nearest in harmony with natural law, and the object and purpose of human life.

"Society as a whole, like a chain with defective links, is no stronger socially, morally, industrially, or politically, than its weakest unit. Hence it becomes the self interest of every individual member to endeavor unselfishly to build up and strengthen the weaker units in every possible way.

"These propositions furnish the only sound basis for a perfect system of political economy—a system which shall afford the greatest amount of good or happiness to all the people. In considering the clearness and startling significance of these truths, we discover the cruel, criminal wrong of any system of competition, based on the old barbaric law of the survival of the fittest, which in its application means the pleasure and happiness of the few at the expense of the toil, pain and misery of the many. In this connection we note that man, in his evolutionary progress, has reached a point where, being mentally and spiritually awakened to a knowledge of the higher purposes of life, he perceives the true effect of environmental conditions, with their good and evil tendencies. He also perceives the cause and the cure. Armed with the talisman of this knowledge, he boldly enters the field of causation and thenceforward becomes a self-directing factor in his own evolution. At this important stage, he clearly comprehends, that the injury of one is the concern of all; that the perfection of all becomes the highest interest of each; that the unprogressive law of the survival of the fittest, is nullified and replaced by the higher law of unselfishness of the individual for the advancement of the race; that the dual nature of man, physical and spiritual, must be considered as inseparable, when dealing with the practical questions of life; that physical life, as the primary school of existence, is ephemeral, while the spiritual is the permanent and enduring; that, consequently, the path of progress for the human soul, lies almost entirely in the realms of the spiritual; that a life on the physical plane, devoted solely to selfishness, dwarfs and chokes the spiritual nature, and becomes a serious bar to unfoldment and progress on the spiritual plane of existence: Finally, that, like the pent up energies of some mighty volcano, the irresistible upward thrust of nature's unfoldment, ever producing and disclosing higher expressions of life, is to find its present outlet through these channels, by the wise use of methods in harmony with the principles stated."

CHAPTER XV.

THE CO-OPERATIVE FARM AS A FACTOR IN SOCIAL EVOLUTION.

"FROM the thorough understanding and appreciation of these principles, by the workers on your model co-operative farm, must come the necessary zeal, the cementing enthusiasm of a mighty purpose which, with ever increasing volume, shall urge them forward to the goal of complete success. As one of the means to insure this success, we must strive to introduce a new era for agriculture, in which co-operative working shall be supplemented and reinforced by co-operative thinking. As applied to farm work, this is a new and untried field which promises grand results.

"In all kinds of productive labor, muscular effort is a mental demonstration! The keener the mentality controlling the muscles, the more satisfactory the work accomplished. The more interested and the healthier and happier the laborer is in his work, the easier it becomes for him to produce superior results. For centuries, farm work has been considered the natural avocation of the ignorant and the illiterate! Strange as it may appear, it seems to have been generally conceded that the typical clodhopper was the ordained farmer! That this perverted idea regarding the requirements of a tiller of the soil, should have maintained its existence for so many ages, is a matter of profound astonishment to every intelligent thinker!"

"Pardon me, Mr. Fenwick," said Fillmore Flagg, "if at this time I quote a case in point from my own state. As late as the year 1897, a Bishop Withington, of Nebraska, speaking of farmers' sons who were struggling for an education, says of them:

"'The farmers' sons—a great many of them—who have absolutely no ability to rise, get a taste of education and follow it up. They will never amount to anything—that is, many of them—and they become dissatisfied to follow in the walk of life that God intended they should, and drift into cities. It is the over-education of those who are not qualified to receive it that fills our cities, while the farms lie idle.'

"This, Mr. Fenwick, is but a sample of many like expressions from the lips of public men, showing the stigma and low estimate which is placed on farmers as a class, by clerical, professional and commercial people. When we consider that farming people form a large majority of the citizens of our republic, a republic whose constitution guarantees equal rights for all; whose chief corner stone from the beginning, has been its admirable system of free education in its public schools; the manifest endeavor of the Bishop and his class, to consign the tillers of the soil to a caste of low order, and to argue that education is for the few and not for the farmer, indicates something radically wrong in our social system that augurs ill for the future of our republic. That the dissatisfaction is widespread and serious, is manifest to all thinkers and observers. To discover the cause and cure, and to speedily apply the remedy for this growing discontent, becomes an imperative duty for all patriotic people. In my experience, the following are some of the most prolific causes:

"The isolation and loneliness of the small farm.

"The long hours of tedious, monotonous toil for both man and woman.

"The constantly increasing competition of large farms, armed with capital and expensive machinery, which tends to reduce the price of farm products.

"The want of proper society, healthful amusements, books, and many other necessary educational facilities.

"The discouraging meagerness of the financial returns for a year of such constant toil.

"These things all tend to destroy the farmer's love for, and pride in, his occupation, until farm work becomes a repulsive drudgery, and he flies to the city for a more congenial employment. Is it then, under the circumstances, any wonder that the farmers' sons should become dissatisfied with the occupation of their birth? That in company with their sisters and sweethearts they should be determined, at all hazards, to escape from the evils of what Bishop Withington terms a 'God-ordained' class of hewers of wood, drawers of water, and tillers of the soil, a class which dooms them and their children to a future of hopeless toil?

"Agriculture forms the basis and support of our national, industrial and commercial success. Therefore it is imperative that agricultural pursuits be made to become the most noble and pleasing of all occupations. How can this be accomplished?

"Surely, co-operative farming, with its improved conditions and methods, is the remedy indicated!"

"Yes, Mr. Flagg," said Fennimore Fenwick, "Co-operative farming is the partial remedy which shall start the healing process, and lead to the discovery of a perfect cure. You have ably stated the evils which make living on small farms so unsatisfactory. You have also made an excellent argument for our work from the text Bishop Withington has so blindly and unthinkingly furnished. It is quite evident that neither he nor his class, have the least conception of the true cause of the discontent they so deeply deplore, It is also equally clear that with all the advantages of superior conditions, with the observation and education of a lifetime, they have so far, utterly failed to understand or appreciate the real object and purpose of human life. They are sorely in need of an object lesson which we must furnish.

"In efforts to slake a natural thirst for knowledge, the brightest minds, the most profound thinkers of the past ten centuries, at the end of lives devoted to study, have declared that the vast domain of knowledge still remained practically an unexplored field. This domain is for coming generations to conquer and possess. It invites the efforts of millions of cooperative thinkers, born and trained for the task. Hence, to me, it is as clear as the noonday sun that the embodiment of more mind by our agricultural people, is a matter of imperative necessity. They should have the leisure and the opportunity to become familiar with all the varied phenomona of nature, through the recorded observations that comprise the different sciences, which describe and explain all phases of surrounding life. Thus equipped, they will be able to discover that they are a living, working, part of nature, which defined, means the combined life of the planet; that they act upon all things about them and are in turn acted upon. A comprehension of these things can come only to the cultivated mind, and the richer its store of facts, the more perfect its grasp and control of surrounding conditions. Therefore mind, as the expression of the soul and body of the dual individual on the physical plane of existence, is EVERYTHING! It controls and molds structure; the body; the people around. All history is but a detailed description of the action of mind.

"The great minds are the dominant thinkers; they sway the multitude, mold public opinion, effect legislation and shape the nation. These dominant minds should come from the people of the soil, as best equipped to discover and proclaim the law of the planet's unfoldment, also best able to conceive and formulate the wise laws which should guide and govern its people. Hence the necessity for our farmers to become thinkers—dominant thinkers.

"What are the best conditions for mind unfoldment?

"As Professor Elmer Gates so wisely says, 'The human body is composed of myriads of living organisms—a co-operative colony of more or less intelligent cells—which respond to the control of the individual Ego through the action of the mind, and to the electrical conditions which flow from the emotions.' Hence the body is an important part of the thinking machine and, therefore, a perfect mind must absolutely be the highest expression of a perfect body. The perfect body needs to be well born. To be well born, is to demand conditions for a perfect motherhood, and the perfect unfoldment of both mother and child together.

"Where can these conditions be found?

"We find them best and most abundant in the rural districts, far from the turmoil and strife, the smoke and poisonous gases of the great city. Surrounded by fields and forests, in the pure air of a broad expanse of country, domed with the blue sky, and flooded with golden sunlight, on the soil of the farm, close to the fostering bosom of our planet mother, Earth. Therefore it must be the distinctive and well defined purpose of our co-operative farm to furnish and perfect these conditions, thus uniting in perfect harmony stirpiculture with agriculture, a union as poetical as it is practical. From these conditions must come a race of dominant thinkers, the exponents and champions of the real objects and purposes of human life.

"With the coming of such a race, comes the beginning of the era of unselfishness, and the end of the present era of selfishnes, the age of gold worship, where greed for gold blights and withers public and private conscience, dominates and corrupts all forms of society, and makes conditions which breed monopolies, caste, tramps. paupers, armies of idle men, strikes, discontent, starvation and revolution!

"Verily, a perfect catalogue of the ways and means by which 'Man's inhumanity to man, makes countless millions mourn!' With the dawn of the unselfish era, comes the demonstration of how man's humanity to man can and will make countless millions rejoice!

"In selecting the people who are to be the active, working members of our co-operative farm, it is a matter of the utmost importance that they should be chosen from a class of persons who are capable of thinking in harmony on religious and political questions, who are already in sympathy with progressive ideas and co-operative work, intelligently alive to its importance and to its advantages, capable of understanding and appreciating that it is not the sole purpose of the organization to make money but also to accomplish a multitude of things besides:

"First and foremost, to ennoble the occupation of their birthright.

"To make farming the most charming and healthful and most desirable of all vocations.

"To make it so remunerative that a reserve fund can be accumulated, sufficiently large to enable its members to purchase the necessary land for an ever increasing series of co-operative farms, for their children and their children's children for generations yet to come.

"To unite stirpiculture so closely with agriculture that a race of perfect children shall be the crowning glory of all the productions of the farm.

"To afford ideal conditions for motherhood and childhood, that all children may be proudly welcomed to a world of loving hearts; that they may be well born, wisely and beautifully unfolded mentally, morally, spiritually and physically; that they may be skillfully taught how to work, to think, to reason, and to comprehend and appreciate the true purposes of life, consequently their duties as true men and women—self-poised and noble, a law unto themselves—capable and fully prepared to enter the walks of life as worthy and honored citizens of an ideal republic.

"That it is to be the province of the farm, by the co-operative thinking of its workers, to develop and increase the fertility and productiveness of the valleys and plains to such an extent that the hills and mountains may be reclothed with beautiful forests of choice trees, of varieties most valued for lumber and timber; also great orchards of the choicest varieties of fruit and nut bearing trees, as a source of future pleasure and profit, at the same time preparing the way for a more complete control of climatic conditions. By the process of shading and protecting the slopes of both hill and mountain by these valuable forests, a magical change for the better is effected. Everywhere a soft, spongy carpet of fallen leaves, ever increasing in thickness, is spread out, moistening and enriching the soil and conserving the waters of the increased rainfall. A thousand living springs of pure, sparkling water make glad the plains and valleys. The evils of flood, erosion and drouth are checked; the climate made more congenial; the value of both hill and mountain, as a source of wealth, increased a thousand fold.

"Aided by the organization of our co-operative association, which makes it possible to treat large tracts of land as a single farm, this great work can be easily and surely accomplished by the earnest and united efforts of a people who, surrounded by conditions of comfort and plenty, are in a suitable mood to plant what their children and coming generations may enjoy.

"As an evidence of man's awakening consciousness of his power, by means of intelligent co-operation, to make conditions that shall protect him and his loved ones from the many calamities which have hitherto beset and overwhelmed human lives, we note the extraordinary work accomplished by the different classes of insurance companies, during the past fifty years. These companies are in fact large bodies of people, incorporated and working co-operatively and systematically together to protect themselves. The success which has followed their efforts in this direction has, for the thinker, a marked significance, pregnant with suggestions for the future. In the co-operative farm, organized and carried forward on lines in harmony with the principles and purposes before stated, this system of insurance, in its simplest, least expensive and most practical form, is to be carried to its fullest extent into all the departments of life. By its wise provisions for the care and protection of the weaker units, it insures its members against loss of employment or wages; against sickness, injury or accident; against poverty, hunger and crime. It insures to all, for themselves and their children, the perpetual right to occupy and till the soil, and thus to secure by short hours of pleasant, attractive labor, the generous return which can be obtained only by the most perfect system of scientific, co-operative farming, armed with abundant capital. In addition, it insures to them all the advantages of birth, health, education, society and amusement which money can buy for the wealthy: more leisure, more opportunities for mental, social, ethical and scientific self-culture. It also insures to the world at large an object lesson which shall demonstrate that the way is open for the poorest farm laborer to secure the same results by joining these progressive co-operative bodies.

"In looking forward to the effect upon society which these combined farms may have, we must consider the numbers and strength of the opposing force which, on every hand, will rise up as a bar to progress. For years, gold, that concentrated essence of selfishness, has been recognized by its worshipers as the crowned king of society, whose crimson banners have borne these suggestive mottoes: 'I am not my brother's keeper! His injuries concern me not!' 'Every man for himself!' 'It is well and good and right that the happiness of the few should be secured at the expense of the misery of the many, for is it not written, "The poor ye have always."?'

"Fortunately, the law of compensation limits and finally crushes the reign of selfishness, causing it to perish by its own efforts to live, which in time destroy the substance upon which it feeds. Hence we may look hopefully to the future. With prophetic eyes we may behold the victorious march of these farm units by companies, battalions, regiments, brigades and divisions, like a vast army of peace, silently spreading, absorbing and conquering the old selfish system, grandly demonstrating the solidarity of human life, and the irresistible force of the combined efforts of thousands of bravely unselfish souls, working and thinking in unison, filled with enthusiasm kindled and inspired by the magnitude and grandeur of the true purposes of life.

"Having thus broadly outlined the scope of the work, with its underlying principles, we may now give attention to the details of the plan for the initial farm. In this I would advise that the enterprise be made to adapt itself, so far as possible, to the present commercial and industrial conditions. That it be an incorporated stock company, limited. That its corporate life be for the longest possible term of years, with the right to renew. That it shall secure and control at least five thousand acres of land, to more readily enable it to dominate the township, as the lowest political unit of the republic; and also to give room for the planting of suitable forests. That its capital stock be limited to one thousand shares, to be divided equally among five hundred co-operators, composed of two hundred and fifty couples or families. That at the end of five years the stock be issued to the subscribers as paid up stock, by cash from the sinking fund, paid in for that purpose. That the stock of a retiring member can be sold only to the treasury of the company, the same to be re-issued to the succeeding member. That in order to avoid friction with the outside commercial world, the stockholders collectively shall sell to themselves individually, at ruling market prices, whatever they may need, the profits to go as a contribution from all to the insurance fund for the aged. That the care of the sick and the injured, and the education of the children, be classed and paid as a legitimate expense of the farm. That the co-operators collectively, pay to themselves individually, a wage sufficiently generous to enable them to purchase what they may desire in the way of furniture, food and and clothing; allowing for a liberal percentage to be devoted to the sinking fund, to pay for the farm, the stock, and also for the additional land that may be secured as future farms for the children. That all other details necessary for the successful carrying out of these plans, be left for a satisfactory solution, to the practical working and co-operative thinking of the members of the farm.

"I wish you, Mr. Flagg, as soon as may be convenient, to make a tour of inspection for the purpose of selecting and purchasing ten of the most available sites for such farms that you can find. From the ten you shall choose the one best adapted to the conditions required for the initial farm.

"After occupation, at the end of five years, these lands are to be sold to the co-operators, at the purchase price, which, in any event, must not exceed the sum of ten dollars per acre. Until the deeds are made to the co-operators, these lands are to be in your custody as sole agent and director.

"In these matters my daughter, Fern, will aid you in every possible way. Many times you will find her advice valuable, therefore when needed, command it withont hesitation. I have an abiding faith that her inspiration will benefit you in many ways in achieving success for the model farm; a matter in which I am greatly interested and to which, as both mortal and spirit, I have for a number of years given close attention and much earnest thought. I now leave the matter to you and to Fern for such thought and discussion as the occasion may demand. I shall be glad at any time to answer questions concerning any particular point. Good night, Mr. Flagg; Good night my daughter."

As Fennimore Fenwick bade them good night, both Fillmore and Fern returned the salutation, and Fern rose from her chair, saying:

"I think, Mr. Flagg, that until now I have never quite understood the broad principles of real unselfishness. In the light of my father's comprehensive statement of the true purpose of human life, they stand forth in bold relief, clear and strong. What a grand incentive they offer, to stir the zeal and enthusiasm of our co-operative workers! All life is affected by them and discloses new meanings. All life seems more precious, more sacred. Yet the task assigned to you, Mr. Flagg, is not an easy one: I foresee many difficulties, but you will overcome all of them. The plan is so thoroughly in harmony with right and justice, so fraught with happiness for the masses, that it must succeed! I trust that you feel encouraged to go forward hopefully with the work?"

"Thanks to Fennimore Fenwick," replied Fillmore Flagg, "I am armed against all obstacles by a new philosophy of life. Its possibilities, as applied to practical work, are beyond computation! His masterly statement of the true theory and purpose of human life, embodies the crystallized wisdom of centuries. I am profoundly impressed with it. Applied to my chosen life work, it demands my best thought, my entire devotion: to co-operative work as exemplified by our proposed model farm, it means unqualified success!

"Pardon me, Miss Fenwick, you have been hard at work, writing rapidly for a long time. You need rest. Let us then postpone further discussion until tomorrow."

"Yes, I think that will be best," replied Fern, "so good night, Mr. Flagg."

"Good night, Miss Fenwick."

CHAPTER XVI.

FILLMORE AND FERN.

For Fillmore Flagg, a never-to-be-forgotten week has passed since the interview with Fennimore Fenwick, noted in our previous chapter. He is still at Fairy Fern Cottage, busy with preparatory work for his coming tour. Momentous events, which have radically changed his life, have followed each other in quick succession. Hours have passed as moments fly, in absorbing interviews with his spirit father and mother. His store of questions in relation to their experiences in spirit life, have all been answered: these answers have in turn suggested many more, until now he is satisfied. For him, the two worlds have been united—the continuity of life beyond the grave has been established as a verity past contradiction. As conscious individuals and loving parents in the realms of spirit life, his father and mother are as real to him as mortals. With each succeeding interview this conviction has grown, until, fully conscious of their loving sympathy and support, he begins to comprehend the connection between life and immortality; the stupendous meaning of immortal life—of never-ending progression—overshadows and dominates all other thoughts. In profound reverence he repeats to himself:

"How noble, how sacred, how wonderful is life! A few years, comparably brief as moments, on the mortal plane of existence, to be followed by an endless Eternity, spent in gleaning wisdom and happiness from the rich fields of infinite progression. By the measure of immortality, who shall attempt to describe or limit the destiny of a human soul? As the epitome of the planet, the universe, and the universal cosmos, it must follow that the human soul is the repository of infinite possibilities. This, then, is the spiritual heritage of all. Sin and suffering, selfishness and greed, crime and vice in the transitory stage of the mortal, might stain and retard his spiritual growth, but they could never destroy the glorious possibilities of the final unfoldment."

This broad conception of the possibilities of human life, here and hereafter, came to Fillmore Flagg as a revelation of the most sacred and marvelous character: in the light of such a revelation, the hideousness of selfishness stood revealed like a grim and warning monster. Now he saw the path of duty plain before him. On the higher, broader plane of unselfishness, he must strive to develop new powers and new aspirations to aid him in making better conditions for a more perfect protection and unfoldment of human life. To satisfy his highest ideal, he must devote himself to this work. The inspiration of the two worlds was upon him! His love for Fern Fenwick, the personification of all that was noble and beautiful, urged him forward; intensified and developed his highest aspirations for good; permeated, glorified and dominated his entire being. Love and life!—the former, the mystery and the crowning glory of the latter.

Hours of self communion, alone in his room, had for Fillmore Flagg a hitherto unknown charm. The crowding memories of the happiest and by far the the most important week of his life, with a tenacity like fever-born visions, passed through and occupied his mind again and yet again. The bright image of Fern Fenwick was the central figure of each event, her grace and beauty was its chief point of interest.

At her unrivaled cottage home he had been the honored guest to whom she had paid her undivided attention. Thanks to her wonderful mediumship, he no longer felt himself an orphan—the gateway of death was also the gateway of life. His father and mother had been restored to him, joined again to his life—his heritage of immortality assured! The truth had been made plain to him that the people of the two worlds were joined by everlasting ties of love and sympathy into the one great flood of humanity, all human beings, all immortal spirits, incarnate, excarnate.

Again, to Fern's mediumship he owed his acquaintance with Fennimore Fenwick, whom he had learned to know, to admire, to love and respect as the highest type of a wise, great and noble man. How fortunate he was in having so many opportunities for learning from such a great master! He prophesied then and there, that the gratitude of coming generations was to bear witness to the power, wisdom and eloquence of Fennimore Fenwick's teachings.

How the memory of all these things swelled the tide of love for Fern Fenwick, in the heart of Fillmore Flagg. How bright and amiable, how gloriously beautiful she was. How kind and gracious she was to him, and what a delightful deference she paid to his opinions! Would he ever again experience another week so full of unalloyed happiness? He had but to close his eyes—a radiant vision of Fern Fenwick was before him, thrilling his heart with hope, urging him forward to the goal of duty. With a sigh he thought of the coming journey. For one blissful week, in the light of her angelic eyes, in the radiance of her loveliness, in the subtle charm of her magnetic presence, he had basked as in the sunshine of paradise: now the hour of parting was approaching, he must not allow himself to be despondent, that would be unmanly; he must hope, wait, and work. Surely his star of destiny augured well for his future. Doubt he could not; doubt he would not! Yes, he would banish all thought of parting. He would think of the work, of its demands, of how Fern had helped him to prepare for it. Oh how proud he was of the peerless girl that had grown so dear to him! As he recalled the many hours they had spent together in discussing the plans of Fennimore Fenwick, as applied to the several stages of development of the model farm, how he had admired and appreciated Fern's brilliant ideas, her pertinent suggestions, her wonderful power to foresee administrative difficulties and to provide most efficiently against them. How well these accomplishments attested the high order of her intellectual training; how perfectly they demonstrated the astuteness of her power of thought, when applied to practical subjects. With such mental and spiritual attributes, supplemented and intensified by the deep inspiration and the awe inspiring majesty of her mediumship, how immeasurably superior she appeared when compared with other women. What problem in life so knotty that she could not solve? With the aid of such a matchless woman, how could he fail in the work before him?

Together Fern and Fillmore had examined many maps for the purpose of deciding on the particular states to be inspected during the coming tour. The great south-west seemed to offer the best field for choosing. The Indian lands, just coming into market, were not to be ignored. They were located in a climate that would promote the growth of a large variety of crops, therefore were especially desirable. Much time was spent by them in going over these important questions very carefully. Fennimore Fenwick, from time to time, had given his opinion on many doubtful points. Now everything was settled. Tomorrow Fillmore Flagg was to start for the rich lands of the great west and south-west, with careful instructions to keep Fern Fenwick informed, by frequent letters, of his progress and whereabouts. Whenever a particular plot of ground was selected, Fern was to send him a certified check for its purchase. This plan was to be followed until all of the desired plots had been secured. The preparatory work on the model farm was then to be commenced.

On the eve of his departure, Fillmore Flagg in reviewing these arrangements, began to percieve that many days must pass before he could hope to see Fern Fenwick again. The intensity of his love for her urged an immediate declaration, that he might know his fate before commencing his long journey;

on the other hand, prudence counselled a more patient waiting and wooing as the only safe and honorable course for him to pursue, as to declare his love at this time would be, under all the circumstances which had made him a guest at the cottage, taking an unfair advantage of the confidence and hospitality of his charming hostess, who had become so inexpressibly dear to him. Yes, he would take up the burden of his work, full of confidence in the wisdom and watchfulness of his guiding star. Hope whispered in his heart: "Fern's destiny is so closely interwoven with thine own, that no fear of the future need disturb thee; in peace and contentment await thou the fulfillment of thy brightest hopes."

Meanwhile, in the heart of Fern Fenwick, the impression left by the events of the week, were marked and apparent even to herself. A change in her regard for Fillmore Flagg was manifest. He was so capable, so loyal to her, and to her interests; and withal so intensely in love with her, that in turn her admiration for him grew apace—in fact she did not attempt to hold it in check. She adored an honest frankness as much as she despised smooth deceit. She knew that Fillmore Flagg was the soul of honor and that she could trust him under all circumstances, else her father would not have chosen him to be her worthy and trusted assistant in the work. In manly beauty he was very near to her ideal; in nobleness of heart, intellectual development and training, he was her equal: therefore it was but natural for her to bestow glances of encouragement on a lover so attractive, so cultured, so unselfish and so ardent. Perhaps she had met her fate! However, before dismissing the subject, she decided at the first opportunity to call the attention of her father and mother to the matter and ask their advice, which would govern her course in the future. She felt that whatever the advice might be, in any event, it would not mar or blight her true happiness.

CHAPTER XVII.

SOLARIS FARM.

ONE year from the time Fillmore Flagg left Fairy Fern Cottage on his trip to the west, we find him at "Solaris Farm," the title chosen for the model or experimental co-operative farm. The location was nearly midway, on one of the through lines of railway which connect St. Louis, the great central city of the Mississippi valley, with the gulf and inland cities of the mammoth state of Texas.

The land was beautifully located, the soil was rich and easy to cultivate. The entire tract was well watered by a fine, clear, swift flowing stream. In extent, the farm comprised ten sections, laying compactly together, and making in all, 6,400 acres of choice land. Nine of the sections formed a perfect square, each of the four sides being three miles in length. The tenth section joined the west line of the south-west section in the square, which made the south line of the farm four miles in length The railroad passed through the farm near the north line of the southern tier of sections, touching on the way an ideal site for the farm village. About four thousand acres of the land was broad, rolling prairie, combined with a large proportion of unusually rich river bottom, both well adapted to the growth of a great variety of crops. The remainder of the farm presented a rough, broken surface, with a soil not so rich, sometimes quite poor and gravelly, but being protected by a great bend in the river, was well covered by a valuable growth of timber. The surface of the roughest ground covered large deposits of lead, zinc, mica and several varieties of choice clay. Numerous bold bluffs contained fine quarries of excellent stone for building purposes, also for an abundant supply of lime and cement. A number of the ridges offered unlimited quantities of gravel and sand. Here and there several rich veins of a very good quality of bituminous coal cropped out.

In making his preliminary examination, the quick eye of Fillmore Flagg soon discovered that this eighteen-hundred-acre tract, of what the owners considered their poorest lands, marred and disfigured by a tangle of undergrowth, a confusion of unsightly rocks, gullies and bluffs; was in reality a treasure, a vast store of choice material for coming needs. When the ten sections, including this broken tract, were offered for the lump sum of thirty two thousand dollars, Fillmore Flagg quickly closed the bargain. He was confident that at last, after many weeks of patient searching, a most desirable site for the initial farm had been secured, at the low average price of five dollars per acre. No wonder he was elated and proud of his achievement! The remaining lands of the township were sparsely settled by about fifty families, generally occupying large ranches.

Acting on Fern Fenwick's advice, as soon as the site of the model farm was chosen, Fillmore Flagg prepared an advertisement for publication in three of the leading spiritual papers, setting forth the purposes of the organization, together with the requirements necessary for membership. The applications which soon followed were so numerous that at the end of the first three months he had been able to complete a very choice selection for the colony. Before the end of the next three months, he had placed them on the farm, prepared for active work.

In the accomplishment of this remarkable feat in so short a time, he had the able assistance of his trusted friends, George and Gertrude Gerrish, who were, from the beginning, most thoroughly in sympathy with him and eager to join him in the work. Fillmore Flagg had known them from childhood and had learned to appreciate them as progressive people of the most pronounced type, who were honest, courageous, and gifted to a high degree with the power to win the love and confidence of all who knew them.

George and Gertrude Gerrish were born and reared on Nebraska farms, near the home of Fillmore Flagg. George was thirty-five; Gertrude, younger by three years. They had been married fifteen years and were noted as a handsome couple, being large, tall, straight and finely formed, with strong, even temperaments. Their only son, Gilbert, was a delicate lad, in his fourteenth year, handsome, spirituelle and intellectual to a remarkable degree. He was a real genius, passionately fond of books, art and music; already an accomplished player on both the piano and violin. Yet withal, he was very reticent, sensitive and shy, on account of his small size and deformed body, the result of spinal trouble caused by a fall while an infant.

The Gerrish family, for the eight years previous,

had resided in St. Louis, where George and Gertrude were employed as teachers. When Fillmore Flagg made them a visit while on his way west from Newburgh, he was both surprised and delighted to find them spiritualists.

They at once became interested in his mission, and his plans for the establishment of a model co-operative farm. At his urgent request, they promised to move at once to the farm, whenever located, in order to be prepared to receive the colonists properly as soon as they should commence to assemble. This promise Fillmore Flagg considered a most extraordinary piece of good fortune, and so it proved.

As a result of this wisely planned co-operative work, at the end of the first six months, a carefully selected, most efficient colony, of five hundred adults and one hundred and fifty children, had been assembled and organized; the business of the incorporation completed; the stock all taken; the officers chosen and a general plan of the work prepared.

George Gerrish was chosen as President of the Solaris Farm Company, Fillmore Flagg was made trustee and general manager. The members of the company were young and strong, accustomed to farm labor, full of enthusiasm for pushing forward the work. They were all wide awake and progressive, quick to perceive and appreciate the importance and advantage of applying co-operative thought and co-operative work to systematic farming on a large scale. They were thoroughly in earnest and equally determined to make the model farm a complete success. With such an army of vigorous, intelligent workers, it was easy to accomplish before the close of the first year, the magical changes which had been effected at the farm.

The land had all been surveyed, examined and tested; the farm carefully subdivided and platted, with a view to keeping a complete record, which should include a debit and credit account with each subdivision. The size and boundaries of these tracts were determined with reference to the capacity of the soil to best produce certain kinds or crops of grains, grasses, vegetables, vines, berries, fruits or trees. The crests of ridges, and all rough, gravelly lands, were set apart for timber, fruit and vineyard culture; the separate areas to be devoted to these three classes were carefully calculated, described and marked on the plat. The number of roads required to connect the various fields and subdivisions with the village, were laid out and made passable by building the necessary bridges.

The site selected for the village was quite near to the railroad, and large enough to give abundant space for future factories, shops, lawns and ornamental pleasure grounds. The whole was graded, well drained and artistically laid out around the four sides of a spacious central square. A large, well constructed freight and passenger station, of Solaris brick, was built and established at the most convenient point on the railroad. In this building were the post office, express office and telegraph office, all in excellent business form and perfect working order.

The manufacture of brick had been one of the first industries developed at the farm. An inexhaustable supply of most excellent clay had been discovered just at the edge of the village site, and speedily connected with it by a short tramway. From this clay the product of Solaris brick proved in every way desirable. In form, color, size and design, they were much superior to ordinary brick. With them, the builder could, in one half the time, with less cement, construct walls that were thick, solid and durable, yet presenting beautiful surfaces both inside and outside. These walls would remain for many years in perfect sanitary condition, kept free from dampness by the dry air circulation, due to the constructive design of the brick. The very fine appearance of the new railroad station, so advertised the beauty and excellence of Solaris brick, that orders from abroad soon came pouring in. To fill these orders without delaying the work on the village buildings, it became necessary to double the size of the brick-making plant; also to increase the number of workers. The unexpected development of such a large and profitable allied industry, at almost the first stage of the preparatory work at the farm, so encouraged Fillmore Flagg and his co-workers, so stimulated and quickened the spirit of inventive genius, that thereafter the efficiency and capacity of the machinery kept pace with the steadily increasing demand for brick, that too without further adding to the working force or to the size of the plant.

A deeper excavation of the clay beds brought to light a much finer class of clays, which proved so excellent for the purposes of manufacturing general pottery, terra cotta ware, drain tiles and sewer pipe, that in connection with the brick works, a factory for making that kind of material was at once put in operation. The tramway was extended a half mile further from the village to reach the newly-opened stone quarries and coal mines, passing on the way large deposits of sand and gravel. By means of the tramway, an abundant supply of all kinds of the necessary materials could be placed on the building site very quickly. The best of stone for the foundations, quantities of brick, lime, sand and cement were at hand, waiting for the builder. All this made possible the swift construction of superior buildings, equipped with all of the modern improvements, including artistic ornamentation.

As a result, before the expiration of the first six months after the arrival of the co-operators, the following buildings had been completed and were ready for use: On the south side of the public square, fronting north; one large mill for grinding flour and feed; one extensive building, large enough to be occupied as a saw mill and planing mill, machine, carpenter, repair and blacksmith shop all combined. On the north side of the square, fronting south; one large three story and basement block of apartment houses, sufficiently capacious to accommodate eight hundred people. The three upper stories were high enough to afford twelve-foot ceilings between the floors. The rooms were large, well lighted, well ventilated, and so arranged on each floor as to offer to every family a parlor, sitting room, dining room, two bed rooms, one bath room, and a kitchen. The basement of the entire block was furnished and fitted to be used as a restaurant, with the necessary dining rooms, kitchens, furnace rooms, store rooms and cellars. The light frame dwellings, located on one of the rear streets, which had given a temporary shelter to the people until the completion of the apartment house, were now utilized as work rooms, seed rooms, assorting rooms, store rooms, and for dairy and apiary purposes. On the west side of the square, fronting east, just across the corner from the apartment house, the well-appointed hall of Education and Amusement was erected. It was three stories high, seventy five feet wide, and one hundred and fifty feet long. The upper story was entirely devoted to the library, assembly and amusement hall, with its large stage, numerous offices and ante rooms. The lower rooms were arranged to be used for the business offices of the farm, the spacious school rooms for its one hundred and fiity children, the printing office and editorial rooms of the press club, and the eleven additional club rooms reserved for the use of the adults. On the same side of the square, fronting eastward and separated from the hall of amusement and education by one hundred feet of space, was the Solaris company store; four stories high, two hundred feet wide, two hundred feet long, built around three sides of a beautifully arranged rose and flower garden. The two lower stories were used to display a large stock of general merchandise, while the upper stories were occupied by the force engaged in the manufacture of general clothing, underwear, and in tailoring and dress making. All of these fine structures were built of Solaris brick, with cut stone foundations; the ornamental brick used in the fronts were especially designed for the purpose and proved wonderfully effective. In every particular the buildings were a credit to the company, being beautifully planned, skillfully constructed, and located with due regard for architectural effect. From the preparation of the stone, the making of the brick, lime and mortar, to the final completion of the buildings, including the making and laying of the sewer pipes, nineteen-twentieths of the total cost was represented by the labor of the co-operators. Of course they were led and taught by a few skilled workmen, directed by Fillmore Flagg, who had prepared the plans. The remarkable success achieved, proved a good lesson in the economics of co-operation, of the utmost significance and value; a lesson which filled the hearts of the members of the company with pride and joy, rivited and clinched their devotion to the model farm and opened their eyes to the possibilities of the future.

Having finished this first series of buildings for immediate use, attention was given to the matter of improving the appearance of the public square. In the center of the broad, smooth green, stood the tall, straight flag-pole; from its top floated the stars and stripes. Eastward from the foot of the flag-staff, and slightly raised above the grassy surface of the smoothly shaven lawn, was spread a living flag in true colors, red, white and blue. This flag was of magnificent proportions, twenty-five feet in width by fifty feet in length, and presented such an effective appearance that it soon became the pride and delight of the farm children, an object of never failing interest, a beautiful living motto which expressed their appreciation of patriotism.

While the building operations were being pushed forward, a carefully selected force of workers had been equally busy in making numerous agricultural improvements. Two thousand acres of virgin soil had been broken up and prepared for planting. One hundred acres of the best of this newly upturned soil, so clean and free from weeds, had been planted with a well selected series of vegetables, capable of producing a remunerative crop of assorted garden seeds. The series included all of the best known varieties with the addition of several new ones. As a result of skillful culture and favorable conditions, a great many tons of choice seeds had been grown, gathered and prepared for market. Large propagating gardens had been fitted and seeded with reference to the future demands of fruit and forestry culture. An abundant supply of all kinds of vegetables for farm use had been grown and stored. Goodly crops of corn, oats and potatoes, grown and harvested. Plenty of hay cut, cured and housed. Pastures, roomy enough to accommodate large herds of horses and cattle, securely enclosed, supplied with water and the proper shelter. Small herds of fine cattle and horses secured and well provided for. These herds were selected chiefly for breeding purposes, while a sufficient number of mules were purchased for the needs of the farm work. The bees in the well stocked apiary had already gathered a fine supply of honey from the wild flowers of the surrounding prairies. The extensive yards and buildings prepared for poultry farming on an unusually large scale, were so well stocked and in such fine condition as to promise large profits at an early day.

In reviewing the work at the close of the first year, which included many important items not yet enumerated, the general results were so satisfactory that the officers and members of the Solaris Farm Company were very much encouraged. Owing to sales of seeds and brick in such considerable quantities, together with the manufacture at the farm of almost every kind of building material, the sum advanced by Fern Fenwick, the patroness, for farm buildings and equipment was less than one-half the amount named in Fillmore Flagg's estimate. The amount required for the coming year would be very much less.

The general plan provided for and embraced the supplementing of agricultural work by a series of allied manufactures, such as naturally grew out of the needs of the farm: carpentering, blacksmithing, machine work and repairing, furniture making, turning, polishing, painting, staining and general wood working and finishing, pattern making, broom and brush making, a factory for spinning rope and cordage, basket and all kinds of osier weaving, brick making, pottery and all kinds of clay or porcelain work; together with many other things that would suggest themselves as time passed and the capacity of the farm was increased by the invention of better machinery and superior methods.

The application of inventive genius on the part of the co-operators to operations at the brick works and pottery, had already proved equal to the demands of any emergency which might arise. The great variety of these added employments would afford a pleasant change from the monotony and routine of ordinary farm work. They could be pursued sometimes for weeks together, when legitimate farm work would be out of season, in this way so greatly increasing the products and profits of the farm, that the bonanza farm of the capitalist, which depended on wheat growing alone for profits, could no longer successfully compete.

After much discussion by the board of management and the officers of the company, it was decided with the unanimous consent of the membership, that eight hours should be considered a days work—six hours for the farm work, with two hours additional to be devoted to such of the manufacturing works as the member might choose. This course proved entirely satisfactory; it soon gave to the farm an able corps of skilled workmen, at the same time augmenting the collective power of the membership to do more effective co-operative thinking for the advancement of the best interests and general welfare of all.

In the matter of wages, a uniform price of three dollars per day was fixed for each member of the company; this amount was diminished by deducting ten per cent for the sinking fund, five per cent for the general service fund, and five cents daily from each member for the special fund. The special fund was for the purposes of education and amusement. After subtracting these deductions, two dollars and fifty cents were left as the net per diem pay of each one. The assessments provided the goodly sum of $54,000 00 annually for the sinking fund, $27,000 00 for the general service fund, and $9,000 00 for the special fund.

The Solaris Farm company was incorporated for ninety-nine years, with a provision for re-incorporation at the expiration of that period. This provision practically made the company a perpetual institution. The stock of the company was capitalized at $250,000 00, and divided into one thousand shares, with a par value of $250 00 each. The number of share holders or subscribers was limited to five hundred adults, about two hundred and fifty couples or families; at the end of five years, two shares of stock were issued to each subscriber, male or female, married or single. This stock, however, could not be issued until $45,000 00 had been paid into the sinking fund. With the issue of the stock, the purchase price of the farm should be paid from the sinking fund to Fillmore Flagg, the trustee, who would then deed the farm to the corporation. Thereafter the company was to maintain a sinking fund amply sufficient to provide such additional farms as the children of its members might need.

In accordance with his instructions from Fennimore Fenwick, the money received in this way by

Fillmore Flagg, was to be held by him as a trust for the purchase of other farms. It was further provided that the Solaris Farm company retained the sole right to purchase all stock which might be offered for sale.

The general service fund was to be used in defraying the expense of stocking, equipping and improving the farm.

It was also determined that settlements made with members, who from any cause might wish to leave the company, should be made on a basis of two dollars and fifty cents per day for the time they had been co-operators, with the return of whatever capital they might have invested plus interest at three per cent per annum; all stock subscribed for to return to the company's treasury.

The general plan further provided for the erection of separate cottages, with small gardens adjoining, for the use and occupancy of such families as might desire them. The apartment house, now completed, had many of its suites of rooms arranged for independent housekeeping, but so far, the members of the company preferred to take their meals at the company restaurant, paying for them the ordinary prices. They also preferred to patronize the laundry, general clothing, tailoring and dress-making departments which were connected with the company store. To prevent any conflict with the commercial interests of the outside world, the restaurant and the company store sold food and goods at the ruling market prices for first-class articles, realizing that it was plainly the policy of the company to keep only the best of everything for sale—the generous profits from all sales to go as a general contribution from the entire membership to the insurance fund for the helpless and the aged. As liberal wages afforded ample means, large purchases were encouraged, and all tendency toward a miserly hoarding was discouraged. It was marked that all the members were quick to appreciate the fact that the more liberal their purchases, the more generously they swelled the fund that was set apart to provide for the needs and happiness of declining years. With each passing month it was observed that this particular feature of insurance continued to grow in popular favor.

To enable the company to dispense with a great deal of expensive bookkeeping, to do business with a small amount of actual cash, and at the same time add another check against the disposition to hoard money; the payment of wages to the members of the company was made in Solaris scrip, good at its face value for all purchases made from the company. Whenever cash was needed by any of the members, an order on the treasurer drawn by the president and approved by the general manager, could easily be obtained for reasonable amounts. On presentation of the order, U. S. legal tenders to the amount specified, would be exchanged for the scrip, dollar for dollar; the treasurer cancelling this scrip by stamping across its face the date of the exchange and the name of the member, retaining the cancelled scrip as his voucher for the disbursement of the money. When scrip was exchanged at the store for goods, it was cancelled in the same way by the manager of the store. The plan seemed to work without friction and gave general satisfaction.

At the beginning of each month an executive committee, composed of three men and three women, was chosen by the members of the company. This committee, with the general manager as chairman, made an order of work for each day and assigned the members to the different kinds of work named in the order. These assignments were always accepted cheerfully. The co-operators without exception and without murmur worked steadily and with zeal for one common result. They were keenly alive to both the importance and the advantages of this new kind of co-operative work, which gave them so many hours of leisure for rest and recreation. With the experience of each passing month, they realized more than ever before that sixteen hours out of the twenty-four so devoted, soon stimulated and reinforced the vital energies to such an extent that active labor seemed really desirable. As a matter of fact, each day they began to look forward eagerly to the six hours of farm work and the two hours additional of skilled labor, as opportunities which gave them refreshing and delightful exercise. Exercise that was necessary to promote health and happiness —exercise which left them with an added relish and brighter mental conditions for the enjoyment of the hours of study and amusement that were to follow. Here again, the wisdom of nature's law of compensation was demonstrated. A grave question of the utmost importance to the progress of mankind was for them forever settled. The discovery had dawned on the minds of these people that labor, no longer a curse, was in reality nature's richest blessing!

Among the more important improvements on the farm which Fillmore Flagg had carefully planned, was the necessary preparatory work on the large propagating gardens, located near the river, not far from the village. In connection with the construction of the village water works, at the time of the grading and sewering of the village grounds, these gardens were furnished with a complete system of irrigating pipes. These, together with the thousands of pots required at a later period, were made in the pottery at the brick works—another product of farm labor. With such a complete control of the necessary moisture, the sprouting process in the long seed beds proved unusually succssful. These beds, which covered several acres of very rich soil, were thickly planted with all kinds of fruit and tree-bearing seeds; together with grape cuttings, mulberries for the silkworm culture, quinces, currants, tea plants, a great variety of berries, a fine selection of ornamental shrubbery, dwarf fruit trees, roses, and many other plants besides. The young plants soon reached a stage of growth where potting became necessary in order to make them strong, well grown, independent young shoots, ready at any time to be transplanted without injury into nursery rows, the vineyard or the berry plots.

To pot the contents of these beds required the labor of many hands, consequently the task furnished a pleasant, congenial employment for a major part of the female co-operaters. A large, well floored, wide roofed shed was constructed just at the edge of the gardens nearest the village. It was wide enough to accommodate two rows of roomy tables, and of a length sufficient for fifty tables in each row. Adjoining the end of the potting shed towards the village, was the storehouse, containing quantities of prepared soil and a large supply of assorted pots. A double track system of narrow tramways passed between the rows of tables, on its way from the storehouse to the different seed beds in all parts of the garden. On this tramway the little cars came from the storehouse to the tables, laden with supplies of pots and prepared soil; these they exchanged for trays of potted plants to be returned to the seed beds. In returning from the gardens on the other track, they brought cargoes of shallow trays filled with little plantlets just lifted from the seed beds. This cargo-bearing process, on the part of the tram cars, continued throughout the day as often as required, making light work for all concerned. To witness the work under the shed as it goes bravely on is a pleasing sight. Let us pause a moment to enjoy it.

At each table are two operators, who may sit or stand while they work. Protected by strong gloves, the deft fingers swiftly fly—the long, double lines of maidens and matrons are as merry as crickets! The buzz of musical chatter, song and story, inspires the work, fitting time with swift pinions and transforming such toil into six hours of fun and frolic!

This class of work proved so charming that a majority of the women preferred it to employment in the apiary, dairy, nursery, school, office, restaurant, or any department of the company store.

With this glimpse of the general development of Solaris Farm, its improvements and its people, during the first year, we discover that Fillmore Flagg has been a very busy man; that his skill, inventive genius, and executive ability have been tried severely; that he has been able to respond to the demands of every occasion. However, such was his confidence in the wisdom of Fern Fenwick, that when he found himself puzzled or in doubt, he relied largely on her advice to suggest some proper solution for each vexing question. He had, from the beginning, furnished her with a complete history of every stage of the development of the farm, along with his weekly reports. At the close of each one he gave a list of topics on which her opinions were solicited; the suggestions in her replies led to such a speedy unraveling of the tangled situations and troublesome questions, that Fillmore Flagg was impressed more than ever, with her excellent judgment and the brilliancy of her genius. His admiration grew; his love grew faster! In his personal letters, transmitting the weekly reports, the expression of these sentiments of admiration and adoration continued to grow in force and fervor until he finally gained courage to request permission to address her as a lover: a lover whose happiness would be largely increased by every effort he might make to put in words the thoughts born of his devotion to her—the one adorable woman in the world, for him.

In her reply, Fern Fenwick frankly stated that she was inclined to consider his request with some degree of favor. That she had sought advice from her parents. That in response her father, Fennimore Fenwick, had expressed himself as convinced of the integrity, honesty, and purity of Fillmore's love for her; but he could not consent to an engagement binding his daughter to marriage, until the unqualified success of the model farm, at the end of the first five years, had demonstrated the worthiness of Fillmore Flagg. After that event, if both continued to desire a marriage engagement, his consent might be considered as assured. Her mother, she said, had repeated and emphasized her father's advice: this advice she felt in duty bound to heed and respect. Therefore, on the conditions named, she was willing to accept him as a lover, with the distinct understanding however, that he must not claim her hand in marriage until after the achievement of the complete success of Solaris Farm.

In the postscript at the close of her letter, Fern adroitly, though perhaps innocently, lighted the torch of hope in the heart of Fillmore Flagg by archly expressing herself as follows: "Henceforth my personal interest in the progress and final success of the model farm will, no doubt, fully equal your own."

This little postscript was a never failing source of comfort and encouragement to Fillmore Flagg. He read it and re-read it again and again: in his ecstacy he caught himself kissing it a dozen times the first week after it reached him. With each reading his hitherto dormant love nature gathered force and intensity. In the throbbing tide of joyful emotions, he was suffused with a strange new happiness. He blushed like a girl as the certainty came home to his heart that at last his love for this beautiful woman was returned. It may be marked as noteworthy that this important letter came to Fillmore Flagg just eight months after his parting with Fern Fenwick at her cottage home on the Hudson. While meditating and luxuriating under the spell of the happy significance of this event, as affecting his future life, he thanked his angel friends for so successfully speeding his wooing. With this assurance he was confident that at last his star of destiny was dominant in the sky of love. Calmly serene, he could now await the approach of whatever trials in life the future might have in store for him. Nothing could shake him from this fortress of love! Nothing could intervene to separate his life from the life of his beloved Fern! With a sigh of contentment, he prepared to devote himself more ambitiously and more industriously than ever before, to the development of Solaris Farm. He wooed every inventive thought; he planned night and day to overcome all obstacles that presented themselves. In his letters to Fern Fenwick, rejoicing in a freedom to express himself without restraint on the limitless theme of his great love for her, he filled page after page with eloquent adoration of his heart's chosen one—his highest ideal of the glorious perfection of womanhood. The effect on Fillmore Flagg of this fervent, all-absorbing love, was most excellent; it broadened and purified his life, eliminating from it all the dross of selfishness. He took a new interest in the lives of every married couple and every pair of lovers on the farm. By persevering effort, tact and skill, he completely won their confidence. He shared their hopes, plans, joys, sorrows, loves and crosses. In all this he never once failed to increase their love for him and their devotion to the farm.

CHAPTER XVIII.

CLUB LIFE AT SOLARIS.

In the work of building up in the minds of the co-operators, an abiding faith in Solaris Farm and its future success, Fillmore Flagg had the able support of George and Gertrude Gerrish. They had proved themselves the right people in the right place! In the schools and nursery Gertrude had become invaluable. Her genial temperament, her fondness for children, the kindly influence of her great mother-heart, with its never failing store of sympathy, patience, tact and skill, all attested that she was a natural teacher whose presence among the children was a perpetual benefaction, while the wonderful store of her personal magnetism brought her the love, respect and obedience of both the old and the young. They instinctively felt her power to make them wiser, better and happier. This was a well merited tribute of praise, worth a king's ransom in gold!

George Gerrish soon became very popular on account of the extraordinary ability he displayed in organizing the members of the farm company into the numerous clubs devised to promote the interests of education, science and amusement. The description which follows will serve to illustrate his skill as an organizer in carrying out the general plan prepared by Fillmore Flagg. In addition they will give a clear idea of the scope and variety of the talent developed, together with a proper conception of the splendid equipment of the farm for the social, educational, ethical and scientific development of its people.

First in order came the Press Club. To it was assigned the duty of editing and publishing the "SOLARIS SENTINEL," a weekly paper devoted to the interests of the farm. It was filled with topics of general interest to the community; themes, essays, poems, personals and social notices contributed by the club members, suggestions and ideas leading to better methods for the care and culture of the farm stock and crops, also as to preparing, the same for market. The range of topics included hints regarding any of the allied manufacturing industries which were carried forward by the farm company. In addition the paper gave full weekly reports from the officers of the different clubs. The literary budget for each week was completed by selections from the general contribution box, a very large one, which was fastened to the outer door of the rooms of the club. Into this box every man, woman and child was invited to drop such written scraps, signed or unsigned, brief or lengthy, as they might be moved to offer for publication. The selections from this box were eagerly read. They often proved surprisingly brilliant, novel or suggestive, frequently disclosing rare literary merit,—altogether constituting the most popular department of the paper. The editorials were carefully prepared and well written. They were usually along lines of co-operative work; its desirability as an encouragement to unselfishness, and also to show how the work might best improve social, industrial and political conditions. The volume and excellence of the reading matter thus produced, was marked by general comment as a matter of astonishment. The unstinted praise which it elicited reflected much credit on the club: therefore to be chosen a member was a coveted honor which was reserved for the meritorious few.

The Dancing Club, in point of popularity, was the most successful of all, and deservedly so. Its membership embraced the entire colony, both old and young who, one and all, seemed to enter into the spirit of the movement with a zealous abandon, a united joyousness, most delightful to behold. The social ties which bound them together, grew and strengthened with the recurrence of each meeting. On two afternoons of each week, the club teachers gave two-hour lessons or drills to all who might desire them. On three evenings of each week, in the large hall of education and amusement, two and one-half hours were devoted to dancing, in which all the members took part. These evening dances proved so fascinating that as a rule very few members were ever noted as being absent. An attack of illness which prevented the attendance of a member, must be desperate indeed. In the matter of general improvement the results were most excellent. To bestow perfect deportment, dignified control of the body and limbs, with an easy, graceful movement on all occasions, there is nothing like dancing. To eliminate the depressing effects of grief, mental or business cares, harassing trials of temper, physical exhaustion, or disturbed spiritual equilibrium, dancing is a remedy of marvelous potency. For the key to the reason why this is true, we are indebted to the wonderful discoveries in psychology and psychurgy made by that able scientist, renowned thinker and brilliant writer, Professor Elmer Gates. The following is a very brief statement of his reasons as to how and why the emotions of the individual affect the vital forces of life:

"The human body is a collection of co-operative cells, more or less intelligent and responsive, therefore an important part of the thinking machine which is acted upon by the superior mind of the brain. The superior mind is in turn reacted upon by the automatic metabolism set up in the cells. Automatic metabolism of the cell, is its ability to carry on within itself the various processes of life that may be necessary to best fit it for the performance of special functions, as a particular part of the co-operative body. Violent emotions of anger, hate, despair and grief, are katabolic, poisonous and harmful; they tear down and destroy life. The poisonous deposits left in the cells by these emotions are called 'katastates.' Laughter and merriment, with all the emotions of pleasure, adoration, worship, love, affection, hope, beauty, etc., are 'anabolic,' or life-preserving. The vital, health-giving deposits left in the cells by these emotions are called 'anastates.' Nature accomplishes her perfect work by beautiful methods. The cells are fed and sustained by the circulation of the blood; they are reached from the smaller branching arteries by a network of minute, thread-like channels, sometimes called 'arterioles.' These arterioles are accompanied by the equally fine wires of the nervous system, closely connected with the brain centers. These wires are electrified by the emotions; they expand the arterioles, and the cells are flooded with an unusual supply of blood; thus they are correspondingly vitalized or poisoned, according to the kind of the dominant emotion, its duration and its intensity."

From the foregoing we readily perceive that the joyful emotions stirred by that poetical trinity, the melody, the rythm and motion of dancing, arouses the circulation so potently that every cell in the body tingles with its superabundance of vitality; both the heart and the brain respond to the invigorating tide, while its precious freight of anastates is vivifying and thrilling every cell. These happifying emotions soon become permanently dominant, the depressing emotions grow weaker, fade away and disappear. The individual is vitalized and rejuvenated! We begin to understand that when properly indulged in, dancing is the most fascinating, healthful and helpful of all the amusements. The Solaris Farm people were both fascinated and benefited by the dancing exercises so generously provided by the club; the growing interest and enthusiasm aroused was a matter of astonishment even to themselves. With the continuation of the club dances, the intensity of the enjoyment and the capacity for it, seemed to increase; this, together with the pleasing memories of bygone dances, seemed to bind them yet more closely to the destinies of Solaris Farm. Strong, straight, lithe figures, happy faces, and eyes shining with the fires of perfect health, gave testimony to the efficacy of music and motion as applied to physical development. With grateful hearts, these happy people realized that this pure font of happiness came to them as the result of unselfish, harmonious co-operation.

The effect on Gilbert Gerrish of this universal spirit of gaiety, was as marked as it was beneficial. On the raised platform at the head of the dancing hall, violin in hand, and surrounded by a chosen few of his friends in the musical club, he seemed to grow in stature as he breathed in the pervading merriment; living a new life, in which his deformity no longer marred his pleasure. Through the association of many months he had grown accustomed to the personal magnetism of the farm people. They were very proud of him and of his many brilliant accomplishments. This all-pervading sentiment of loving pride came to him as a benediction, which his refined, sensitive nature graciously absorbed. His shyness and reticence disappeared; his face glowed with the flush of happiness; his beautiful eyes shone with the fires of a new inspiration. With the hand of a master he swept the strings with a bow of magic; new strains of sweet, thrilling music stirred the dancers and moved them as one mass to the throbbing rythm of the intoxicating melody: a melody so charming that none could resist. Filled with the power of a new grace and dignity at such moments, Gilbert Gerrish felt a keen triumph in his ability to stir the emotional natures of these people whom he loved; to inspire them to better deeds and to nobler lives. They, in turn, recognized and paid willing homage to a noble soul, a great genius, whose power to sway and control them was not in the least deflected or dimmed by a thought of his deformed body. Under the mystic spell of divine music, which appeals to the highest aspirations of the human heart; which calls forth the hidden forces of the soul: they came in such perfect rapport with him in his inner life, that they sensed with soulful eyes the strong, radiant, symmetrical spirit shining through the defects and barriers of a fleshly prison. Thus transfigured, they saw him, not as he appeared to ordinary mortals, but as he really was, To these people of Solaris, this transfiguration was lasting. Very soon they came to regard him as a talisman of good fortune—the mascot of the farm.

The Photographic Club, organized by George Gerrish soon after the press club with the intention of making it the nucleus of a future art club, proved a surprising success at an early stage of its existence. Very soon after active work began, fifty members had been enrolled, In discussing with the executive committee a general plan of formation, Fillmore Flagg remarked that he felt very sure the club would soon prove a valuable aid to the farm in the direction of furnishing attractive illustrations of the farm itself, its products, stock, fruits and flowers, to be used as advertisements. With this in view, he made arrangements to provide suitable rooms, large, well lighted and fitted for the work, in connection with the construction of an isolated building, made as nearly fire-proof as possible which, when finished, was to be devoted mainly to the needs of farm experiments in the department of agricultural chemistry. The completed rooms, with a large lot of cameras of various sizes, together with an abundant supply of photographic material, were placed at the disposal of the working members of the club. These things were rightly considered a necessary part of an educational outfit. Fillmore Flagg and George Gerrish both were skillful photographers: with the wise guidance of two such able teachers, the class soon began to produce creditable work. After the expiration of a fixed period, in compliance with an imperative club rule, each member was obliged to complete all work from start to finish without assistance. This would give scope and opportunity for expressions of spontaniety and inventive genius in the individual treatment of the work, which might tend to the evolution of superior methods. It was clearly an advantage for the members to be able to say truthfully that photographs produced under such requirements were actually the results of their own individual handiwork; from focusing the object, timing the exposure of the plate, on through the various stages of developing, toning, printing and mounting, up to the final process of polishing the finished picture. At the end of each month the members individually were required to submit twelve finished photographs to the inspection of a committee of five. This committee was composed of two ladies and two gentlemen in addition to Fillmore Flagg, who was the chairman. From this collection of twelve lot pictures, representing the finest work of the club, the committee selected four photographs from each lot, which were chosen to become a part of the farm exhibit to be displayed on the walls of the library, hall of education and the school-rooms. This monthly award for meritorious work acted as a wonderful stimulus to all the club members, so increasing their ambition, industry and artistic invention, that an ever increasing number of delightful surprises followed each monthly examination. In considering the selections as a class, the extent and variety of the subjects treated covered a wide range. Among them we may name the general and special views of the farm, its buildings, fields of grain, corn, cotton and broom corn; bits of forest, meadow or brookside landscapes; specimens of the different vegetables and garden products; interior views of the different buildings; photographs of groups and of individual members of the company; pictures of manufactured articles, tableware, ornamental brick and tile work, and general pottery; a great variety of cabinet work, furniture and willow ware; splendid photographs of horses, mules, cattle, sheep, hogs and poultry, also wild animals and birds, singly and in groups; views of trees, streams, roads, bridges and railroad trains; enlarged photographs of the insect enemies of farm products; others of the birds which prey upon such insects; artistic views of seed beds, nursery rows, potting sheds, brick and pottery works—in fact, pictures of every possible aspect of the agricultural and manufacturing industries on the farm. Taken together, this collection presented a most interesting series for the school rooms, which proved an object lesson of great value to both pupil and teacher. The landscapes were especially excellent in giving correct ideas of distance values in perspective drawing.

As time passed, the inventive genius of the club members began to crop out in the repair shop, where they not infrequently, and sometimes much to their surprise, found themselves able to construct better and cheaper instruments, lenses and attachments than they were able to buy. With these improvements they soon achieved success in color photography. Later this led to making magnificently colored slides for stereopticon, kinetescope and biograph exhibits, which soon attracted wide attention and were in such demand that a large trade resulted. In this way another exceedingly profitable allied industry was added to the now famous Solaris Farm.

CHAPTER XIX.

FENWICK HALL.

IN THE infancy of this Republic, when its government was looking about for a permanent home, Gen. Washington was moved to found and lay out the City of Washington as its Capitol. With a marvelous prescience he foresaw the coming needs and future greatness of the newly-united states. Impressed with visions of the glorious destiny awaiting his beloved people, his cherished republic, he wisely concluded to provide generously for the growth of a magnificent city which, a century later, should reflect credit as the capital of a mighty nation. Careless of the gibes and sneers of many of his most intimate friends, Washington, the far-seeing statesman, the invincible soldier, deliberately planned, platted and surveyed through the wilderness of forest at that time covering the great triangular basin lying between the Heights of Columbia and the waters of the Potomac and Anacostia rivers; such a bewildering array of broad streets, wide avenues, and roomy public parks, as would be ample and suitable for a brilliant city like Paris, (whose system of streets he had taken as a model,) at least sufficient for the wants of a population of a half million. The dawn of the twentieth century saw a complete realization of General Washington's brightest hopes, a verification of his prophetic visions. The wand of progress had transformed the straggling village of "magnificent distances," into the most royally beautiful city on the continent. A city which had become the pride and delight of one hundred millions of free people, who individually felt a personal interest in the vastness, the beauty and the imposing grandeur of its magnificent public buildings, which represented the crowning loveliness of architectural design, the highest artistic expression of American genius; altogether most perfectly and fittingly adorning the unrivaled capitol city of the most progressive, powerful, and meritoriously dominant republic on the face of the planet! To this Mecca of republics, as the social and political center of the western hemisphere, came the great thinkers, scientists, artists, orators and statesmen of the world.

Commandingly situated on Columbia Heights, overlooking this surpassingly beautiful city, was Fenwick Hall, the home of Fern Fenwick. The Hall was a large quadrangular structure of imposing appearance, erected in the center of spacious grounds, most charmingly laid out, with a rare combination of lawn, flowers and shrubbery. The material used in its construction was Seneca sandstone, in color a rich dark red, and was trimmed with a pale mottled green stone, quite as beautiful as serpentine. The effect of the combination was as harmonious as it was ornamental. The main building was four full stories in height above the deep basement. It was made more conspicuous and more picturesque by the four octagonal towers, one-half of which projected from each corner of the building. These beautiful towers of a uniform size, rose thirty feet above the roof of the building itself. The basement and towers were of rough green stone; the caps and sills of the long, deep windows, together with the arcade, were of green stone, beautifully carved and polished. The arcade, which served both as a covered way, and a portico over the main entrance, was at once artistic and unique. It was formed by a picturesque combination of four Moorish arches. These arches were uniformly twenty-five feet in height and twenty-five feet in width: the openings of the double arch were placed in front with the single openings at either side. By this arrangement the beauty of the entire structure was greatly enhanced, while a very appropriate entrance to Fenwick Hall was the result.

At the rear of the grounds, on a line with the center of the mansion, were the roomy stables. They were built of rough Seneca sandstone. Like Swiss cottages, they were made more beautiful by a profusion of richly colored slates which covered the broad, steep roof and the wide eaves. Between the mansion and the stables, on the same line, twenty-five feet distant from the former, was the pretty two story building, of the same material, devoted to the kitchen, the heating and the lighting plants. Both buildings were connected with each other and with the main building by a long colonnade of harmonious proportions; its heavy cornice, narrow, steep roof, and long double line of slender supporting pillars, were all of the same red stone. The color effects offered by the lovely contrast between the velvety green of the broad, smoothly shaven lawns and the rich reds of the Seneca stone, were simply delightful! Architecturally considered, the combined effect of the group of buildings, arcade and colonnade, was as artistic as it was excellent. Under the arcade, just inside the double arch, a broad flight of stone steps led up to the heavy oak doors opening into the wide hall on the main floor. This hall was remarkable for its unusual size; it was thirty feet wide and of a proportionate height, fifteen feet from floor to ceiling. In connection with a cross hall twenty feet in width, it served to divide the entire space on this floor—one hundred and sixty feet by ninety—into four very large rooms; the two parlors, the library, and the dining room: each one thirty feet in width by seventy feet in length, with fifteen foot ceilings.

The grand proportions of these magnificent rooms and stately halls, excited universal admiration; they impressed the beholder with a dominant idea of the spacious luxury which marked the interior appointments of Fenwick Hall. In the center of the main hall, thirty feet from the front entrance, began the flight of the grand stairway. The general design of this stairway was boldly unique. It was in harmony with the scale of magnificence which characterized the halls and parlors. In three long flights of twenty-five steps each, it rose to the fourth floor. Counting the fifteen-foot landings on the second and third floors, it was practically one structure with a generous breadth of fifteen feet. It was built of the same material—American mahogany—with casings, cornices, banisters and newels of the same pattern and finish, all highly polished and rich with ornamental carving. The beautiful color effects of the polished mahogany, were brought out more vividly by the pale neutral tint of the heavy velvet carpet, which covered the stairs and landings. As an illustration of the great space occupied by this grand stairway of such ideal proportions, each one of its seventy-five broad steps would afford a comfortable seat for eight persons—a goodly company of six hundred, all told. This royal trinity of stairways ranked as the distinguishing feature of the mansion. They gave it an air of stately elegance, tempered with the glow and warmth of a generous hospitality.

The halls on the second and third floors were counterparts of the main hall in size and style. The hall on the fourth floor was fifty feet wide by one hundred and sixty feet long. It was arranged to be used as a ball room, or for concerts, lectures, operas and theatricals. For such events, it would comfortably seat an audience of one thousand people. The roomy stage was furnished with the latest and most approved appliances; it was also equipped with a remarkable series of twelve drop curtains for the lectures. Number one of the series, was a twelve by twenty-four foot map of the United States, including Alaska, Hawaii, Porto Rico and other territorial possessions. This map was accurately drawn to a large scale, it was artistically colored and marked in such a way as to show at a glance the boundaries of original territory; the ceded territory, the date of cession, and from whom acquired; the dividing lines between states and between counties; the location of all cities and towns having a population of one thousand or over; the principal state and county roads, all railroads, lakes, rivers, mountains, public parks, valuable forests, arid lands, irrigable lands, mineral deposits; all noted mines of coal, iron, gold, silver, copper, etc., together with a great variety of important items: all of which proved exceedingly valuable as an added means by which to illustrate in an interesting and comprehensive way, lectures on geographical, geological and historical subjects, together with lectures on the natural wealth and resources of our country; its manufacturing, mining, commercial and agricultural interests, with a great number of kindred topics as well. The second curtain was uniform in size with the first and with the entire series. On the same large scale, it gave a magnificent illustration of the solar system. The background was a pale bluish gray. The sun appeared as the central figure, surrounded by the planets in their orbits, carefully drawn as to comparative size and position. The whole map was colored with exquisite taste in perfect harmony with the beautiful sky effects of the background. The skillful work of the map maker proved especially strong in furnishing a lesson of wholesome humility for the over-proud denizens of the little planet Earth who, puffed up with much vanity, have for ages proclaimed the Earth as the pivotal center of all creation. The third curtain was simply a heavy, plain white one, perfectly fitted for the display of stereopticon views, and more especially for the moving panoramic views of the kinetescope, the vitascope and the biograph, which have proved such attractive and entertaining aids to the general lecturer, dealing with any special subject capable of such profuse illustration. The remaining nine curtains were devoted to outline maps of the world, and to illustrated object-lessons in the most important and interesting departments of nature.

The side walls of this remarkable hall were wainscoted in polished hard wood, for a distance of five feet above the floor: the remaining wall space was divided into large ornamental panels, with beautifully scrolled historical borders. In these panels were painted, one in each, large maps of the States and Territories, which were drawn to uniform scale, minutely accurate, with every post office, post road, wagon road or cycle path plainly marked. In addition, at least twice the number of details usual to large maps showing counties and townships, were carefully noted. The effect of this unique educational system of ornamentation was as interesting as it was fascinating. In harmony with this idea, the entire length of the broad ceiling overhead was painted a pale blue; it was divided into two large panels with ornate borders; each panel was dotted with stars and planets in such a methodical way as to form a complete astronomical map of the visible heavens, both northern and southern hemispheres. This, with several of the large drop curtains, served as adjuncts to the well equipped observatory which was located in one of the large towers at the rear of the mansion.

On the main floor, on each side of the front hall, were the two grand parlors, whose exact dimensions have been stated heretofore. They were carpeted and furnished with all the art and luxury that skill could devise, or wealth could procure. Two wide archways of Moorish style and majestic proportions, opened from each parlor into the main hall. The chiof adornments which marked these fine parlors as unapproachably superb, were two immense mirrors, alike in every way, mounted in heavy frames, rich with leaf gold. They occupied the entire wall space at the rear end of these enchanting saloons of artistic luxury. When distinguished groups of brave men and beautiful women were assembled here, the magical effect of these mirrors in reproducing the brilliant company as one magnificently framed panoramic picture, was ever the source of perpetual admiration and delight. On such occasions the thirty feet of the main hall in front of the stairway, served as the third or reception parlor. The grand stairway shone resplendent as one magnificent centerpiece of loveliness. Up the long flight on either side, it was banked by a wealth of potted flowers, ferns and palms, festooned with wreaths of lovely smilax. Just in front of this unrivaled background of beauty, standing alone upon the movable reception platform, which was merely a small circular extension of the first step of the grand stairway, the charming young hostess of Fenwick Hall, with the grace and courtesy of a born princess, gave a greeting of welcome to her delighted guests, or dismissed them with a gracious smile as they entered or retired.

The library, in the rear of the parlor at the left of the main hall and separated from it by the cross hall, was an exceedingly imposing and attractive room. With its quiet array of costly appointments, it seemed to possess some hidden charm. Its mahogany shelves were laden with a rare collection of choice books, elegantly bound, skillfully arranged and classified. The assortment of scientific books was a remarkably large one. Marble statues, and exquisitely painted portraits of a host of famous authors and artists, whose works had enriched the literature of the world, fittingly adorned this ideal realm of drowsy quiet, where both lore and luxury reigned supreme.

The dining room was uniform in size with the parlors and the library. Its walls and ceiling were frescoed with groups of grceful figures, which represented the merry sprites of pleasure in carnivals of feasting, song and dancing. Each figure was a carefully studied type of beauty; each group a perfect expression of grace and gaiety. Studied singly or as parts of the entire composition, they were exquisite as works of art, charming the attention of the beholder with a bewildering fascination. The floor was one vast mosiac of superbly colored tiles. The heavy mahogany tables and sideboards were glittering with their costly equipments of shining silver, sparkling cut glass, and rare, translucent china. Large oval mirrors in heavy carved frames, duplicated the lovely adornments of this brilliant room from a dozen points of vantage. The dazzling effect of this home of the feast, was intensified by cascades of light from the two unrivaled chandeliers. They supported a great number of slender bulbs containing the electric lights, which were araged in the form of a mass of drooping fern leaves, rising like a pyramid of soft radiance, into the perfect shape of two superb fountains. Tiny streams of short prisms, clear, flashing, crystal, pendant and vibrating, formed the tip of each fern leaf. This skillful combination seemed to complete the startling illusion of this rare vision of loveliness, until one could almost hear the musical tinkle of falling water.

The three halls on the main, second and third floors, were really galleries of art "par excellence," they were so profusely adorned with choice collections of photographs, etchings, water colors, paintings and statuary. On entering the main hall, two very large paintings of extraordinary significance and rare merit claimed instant admiration. Companion pictures, each with a canopy and background of crossed American flags, from whose voluminous folds shone the blazing glory of color in the matchless beauty of the stars and stripes. In each picture under these flags, the dominant spirit of the republic breathed in the noble figures so exquisitely painted; typifying in the one on the right, the Goddess of Liberty watching over the destiny of the republic. In the one on the left, Liberty with her torch lighting the world. So perfectly did the painter's art portray the "Spirit of '76," that a new tide of patriotic devotion to the republic and its glorious flag, swelled the hearts of all who saw these justly famous pictures.

The well lighted, well ventilated rooms in the basement were used as store rooms, a suitable number being set apart for the servants, as dressing rooms, dining room and sitting room. In a large bay window extension at the rear of the main hall, a sumptously furnished elevator connected the basement with all of the halls, the roof and the towers. The rooms on the second and third floors were arranged in suites of three, reception, sleeping and bath. In size, fittings and furnishings, they were models of comfort and luxury.

The four octagonal tower rooms were uniformly twenty-five feet in diameter, with lofty dome ceilings. The right front tower was occupied by Fern Fenwick as her private study and work room. It was fitted and furnished much the same as the library. The left front tower was arranged as a seance room for spiritual manifestations, and more especially for the different phases of mediumship possessed by Mrs. Bainbridge, including materialization. As before stated, the right hand tower at the rear was perfectly equipped as an observatory, while the rooms under it were devoted to the demonstration of kindred sciences. The left tower at the rear was furnished and arranged as a laboratory. The rooms under it were set apart for experiment and demonstrations in chemistry, metallurgy, photography and several other sciences of like nature.

An able corps of carefully trained servants, under the direction of Mrs. Bainbridge, the housekeeper, made it easy to keep this remarkable establishment in perfect order. One and all, these model servants were devoted to their lovely young mistress, and this devotion was based on their keen appreciation of her noble ideas in regard to the true purpose of human life, to her high estimation of its sacredness. They were eager to serve her faithfully and well for less than ordinary wages, contented and confident in the knowledge that, in accordance with her clear sense of justice, they were sure of being retired on half pay after having reached the age of fifty-five. This brief description of the exterior and interior of Fenwick Hall, its equipment, its lovely mistress and its people, will but faintly suggest its extraordinary possibilities as a potent factor in the upper circles of Washington life. Almost three years have passed since the transition of Fennimore Fenwick, which left his only daughter, Fern Fenwick, as the sole heir to his vast wealth. With the exception of three months each summer, spent at Fairy Fern Cottage, or some mountain resort near it, she had remained quietly at Fenwick Hall, busily engaged in rebuilding and refitting it. Meanwhile under the instruction of able teachers, she had been hard at work in efforts to supplement her excellent collegiate education with a better knowledge of history and by a more complete mastery of the subtle secrets of the higher sciences, as exponents of the powers, properties and purposes of the inherent forces belonging to the various departments of Nature's vast domain.

After much deliberation she had undertaken this work to enable her to wisely prepare and plan for a life work in harmony with her lofty ideas on the subject—ideas which had been slowly ripening in her mind for many months. Having passed the ordeal of this severe post graduate course of general study, she felt herself prepared to commence the work contemplated by her general plan, which embraced a skillful use of the great educational and social advantages of Fenwick Hall, in her endeavors to bring to the leading minds of the political and social circles of Washington a clear conception of the importance and significance of the real purpose of human life; with a view to reforming ethical, social, industrial and political organizations on the true basis of the unselfishness of the individual for the advancement of the race; thus bringing these organizations into exact and co-operative harmony with the object and purpose of the existence of the planet. Systems so organized, would then be in line with a true conception of the functions of an ideal republic—a government for the people, of the people and by the people; conducted for the benefit, protection and development of all the people. With the world organized into families of such republics, the advent of the millennium could be predicted, and the advancement of the race to the point of perfection would be insured.

CHAPTER XX.

THE BEGINNING OF A NEW ERA.

FROM A careful review of her historical studies, Fern Fenwick came to the conclusion that the competitive system was responsible for a majority of the evils which had so retarded the world's progress. She discovered that this same system was the father of a conscienceless commercial spirit which had existed for many centuries as the basis of all social organization. That as such, it was a constant menace to all good society; the embodiment of a cruel selfishness of a savage type, which insisted that might makes right—that the strong should thrive by preying upon the weak. In this position it boldly denied the immortality of the soul, so far as the weaker workers were concerned. Therefore the cheap lives of these poor people had no claim to be considered as sacred, because they represented so many human souls. In the absence of any practical or effective protest from the religions of the world, this monstrous system of selfishness had in all these years, grown unchecked and unmolested in its methods of cruel greed. From the shadows and gloom of these threatening conditions, existing so manifestly in direct violation of all progressive law, came a demand that the negative belief in the immortality of the soul, be speedily replaced by a positive knowledge of it. A knowledge sustained and supported by practical demonstrations, through the action of natural law, whose manifestations and demonstrations should be so direct and indisputable as to appeal convincingly to the hard headed thinkers, who as a class, seemed to represent a materialistic element that threatened to overthrow all belief in immortality.

In answer to this demand, about the beginning of the last half of the nineteenth century, there happened an event of the utmost importance, potent with promise for the mighty spiritual unfoldment and general advancement of the people of the twentieth century.

In the humble home of the Fox family, at the little village of Hydesville, near Rochester, New York, by the co-operative efforts of mortals and spirits, there was constructed and established a line of communication between the two worlds—the mortal and the spiritual. Two little children, the Fox girls, were the mediums, a combination of operator and electric battery—or, in other words the necessary instruments for successful spiritual telegraphy. In this obscure home of the poor and lowly, in a quiet way, unheralded and unannounced, there came to the world a knowledge of the existence of one of nature's grandest laws, the law of mediumship; thereafter the way was open, on the physical plane of existence, for an unlimited series of practical demonstrations of the immortality of the human soul: the continuity of conscious life was substantiated by an endless variety of proofs of the most convincing character.

With this solution, of the destiny of the human soul as an immortal and imperishable entity, came the solid ground on which to build a permanent foundation for a social and industrial organization, on a basis of unselfish, harmonious co-operation in perfect accord with planetary evolution, and the real object and purpose of human life.

This strong combination of the working factors of

the problem, suggested to the mind of Fern Fenwick the importance of first attempting to interest the minds of the people she wished to control, in the question of immortality as a natural fact that followed the dual nature of all human life, as a result of planetary evolution. Once interested, she could then convince them of the immortality of the soul, as a conscious, imperishable entity, by practical demonstrations through the law of mediumship.

These demonstrations would make it clear to them that life on the physical plane of existence is transitory and ephemeral; somewhat in the nature of a very brief period of primary experiences; that life on the spiritual plane of existence is permanent and enduring; that therefore the pathway of progress for the human soul must be almost entirely within the realms of the world of spirit; that this great truth should have careful consideration when dealing with questions affecting human lives; that the dominant immortal spirit of the dual individual possesses a corporeal body, or mortal form, as a crude outward expression of the indwelling spirit in its earthly existence; that this mortal form enfolds all the possibilities of a life of eternal progression for the Ego or spirit as a conscious identity on the spiritual plane of existence; that the change called death is a natural one, to be approached calmly without a fear; that it is really a new birth, which does not disturb the continuity of life.

Once convinced of the verity of these great truths, all lovers of humanity, all progressive people, all earnest thinkers, would readily understand and appreciate the sacredness of human life, as the flower and fruit of the planet—its highest expression; they would then be prepared to co-operate with any progressive movement for the advancement of the race.

To make the necessary conditions for the accomplishment of this great work was the grand purpose of Fern Fenwick's Washington life. With this purpose in view, Fenwick Hall had been especially fitted and equipped. For this she had cultivated a large circle of acquaintances among the fashionable leaders of the best society of the Capital City. Caring but little for the ceaseless round of soul-wearying social functions which so completely absorbed these people; yet filled with a determination to win them to a higher life, she bore herself bravely through the season which proved one long procession of social triumphs. Inspired by the intensity of a grand purpose; endowed with a clear, musical voice, perfect health, youth and beauty, combined with a charmingly irresistible personal magnetism; armed with the quiet dignity of perfect self-control, and the genius of her brilliant mind, so broadly cultured; an adept in psychic lore; an entertaining and eloquent conversationalist, our heroine created a profound sensation in the most select circles of the social world. Everywhere she was the center of attraction, surrounded by admiring throngs of cultured people, representing wealth and leisure, who hastened to pay homage to her as a Twentieth Century society goddess, whose wand of magic controlled millions of money. In the homes of the exclusive few, she was hailed as a thrice welcome guest; celebrities, ranking high as statesmen, soldiers, poets, artists, anthors, representative professional men and leading men of business, were completely charmed and curiously fascinated by this new queen of the social realm, and vied with each other in eager efforts to win her favor and perhaps her friendship, in the hope of gaining admittance to the very limited circle of fortunate people who were the recipients of invitations to the famous dinners, receptions and entertainments at Fenwick Hall. These people instinctively felt the attractive power of some silent, mysterious force, some high motive, which, combined with dazzling beauty and brilliant genius, drew them to her side, without the wish or power to resist.

This phenomenal wave of popularity continued to increase until a choice of the best people in every branch of the social world, was at the command of this new leader of the exclusive set; they were ready to assist in carrying forward any progressive movement she might choose, by her championship to make the fashion. However, this universal willingness to follow her leadership, seemed based on a firm conviction in some way unconsciously established in the minds of her devotees, that all of Fern Fenwick's plans and purposes were for the good of humanity, wisely guided by a skill and judgment most remarkably rare—apparently far beyond her years! The whole situation was a complex problem they could not analyze: they did not even try!

With the advent of modern spiritualism in 1848, came the first opportunity to bring woman forward as a teacher and leader in the great work of elevating and spiritualizing the masses. As a heritage from her sister oracles, who spake in the mystic temples of the ancient past, the modern woman was endowed with the divinity of a rarely sensitive and highly refined spiritual organization. By virtue of this endowment, she speedily demonstrated her peculiar fitness for this new mission. Her eloquence and inspiration charmed the multitude from a thousand rostrums. Her work in this new field was so startlingly brilliant, important and successful as to attract the attention of the whole civilized world; affording a remarkable object lesson which demonstrated her possession, as the mouth-piece of inspiration, of a wonderful magnetic power to sway the people; to enthuse, interest and educate them up to higher mental, moral and spiritual conditions; by making them aware of the vast import of the true purpose of human life; by helping them to realize to a limited degree, the significance of immortality, their individual responsibility in relation to the universe, as important factors in the evolutionary advancement of the race toward the millennium of its final destiny.

These inspired teachings touched a responsive chord in the hearts of all womankind as they began, dimly at first, to perceive the all-pervading force and rythm of the dominant key-note to the evolution of the race, which in thunder tones ever proclaims the mighty truth, that all progress of the race depends entirely upon the elevation, education and refinement achieved by woman. They also began to understand something of the glorious possibilities of a perfected womanhood, as a regenerator of mankind. A magnificent array of future victories for woman's work loomed up before them as a command to awake; to prepare for the coming dawn of the twentieth century—the beginning of a new cycle in the life of the planet; the commencement of woman's golden era! To woman the command was imperative that she must strive for more wisdom, for more light on her holy mission as the evangel of evolving life; that she might reach a higher consciousness of her individual responsibility as the keeper and guardian of the sacred temple of human life—a temple in which is ever repeated the evolution, ontogeny, and phylogeny of the race; where, by this most mysteriously beautiful of all processes, there is constantly being welded together the planetary growth, physical, mental and psychical experiences of ages upon ages in the past; with the higher, purer, better and more spiritual possibilities of the race in its planetary progress for uncounted ages yet to come.

From this general awakening there followed—for the purpose of securing that practical education of training, which actual contact and individual experience alone can confer—a vigorous effort on the part of the brightest and most progresive women of the Nineteenth Century, to enter, singly and as organizations, into all the activities of life. Hampered by the blinding predjudice of a long line of centuries; many of these earlier organizations, as might have been foreseen, were unsparingly criticised as exhibitions of ill-directed foolishness, altogether crude, unprogressive and unsatisfactory. Nevertheless, the dominant spirit of courageous and persistent effort, combined with high purpose and pure motive, soon won the approval of the better classes and accomplished a marked improvement in both work and method. This rapid improvement pointed unerringly to future achievment of that success shown in the conditions which prevailed at the close of the century, whereby woman was very generally recognized as a necessary and successful co-worker in all the suitable employments of life.

Fern Fenwick, in full sympathy with the movement, was alive to the demands of the situation. With the purpose of concentrating the efforts of all the women's organizations which held their annual conventions in Washington, into one channel, leading to perfect motherhood, as the result of woman's social and financial independence; she identified herself with them as a generous contributor. Soon she became the friend and trusted adviser of all of the leaders. She placed Fenwick Hall at their disposal, for use as a general headquarters. In this way, a wise direction of the combined women's movement into a united work along lines in harmony with planetary evolution for the perfection of the race, became an integral part of Fern Fenwick's broad plan for a life work.

By the end of Fillmore Flagg's first year at Solaris Farm, Fern Fenwick had matured her plans for her own peculiar work. Much to her satisfaction, the necessary conditions had been created, the whole movement organized and well in hand. Fillmore's work for the education and elevation of the agricultural classes, had given her energy and inspiration to accomplish a similar and co-operative work among people of wealth and leisure, who, ignorant of the true object and purpose of life, were unwittingly wasting precious years in leading indolent and aimless lives, by lending themselves body and soul to the care and canker of the fashionable game of killing time. One year's experience had taught her that the task was a difficult one, to accomplish which required time, patience and perseverance, reinforced by courage, skill and tact.

CHAPTER XXI.

HIS WOOING PROSPERS WHILE OUR HERO ENJOYS HIS FIRST VACATION.

FERN FENWICK'S interest in the experimental farm was intense. She read with eagerness the weekly reports from Fillmore Flagg, which were accompanied by such charmingly ardent love letters. She was very proud of the success he had achieved in two short years. She blushed as she thought how dear to her he had become in those busy months which swiftly passed. How much she should miss him and his fascinating love letters, if by evil chance anything should should happen to take him away from her! She could not contemplate such a possibility without a shudder. Now that her studies were finished and her plans perfected, why not send for him to come to Fenwick Hall for a week's vacation? He had certainly earned the privilege which he would prize so much. The opportunity to personally compare notes and exchange suggestions would no doubt prove helpful to the farm work and to her own. She longed for the confidential companionship of some one who was in perfect sympathy with her, who could understand her work, and appreciate her motives in carrying it forward; some one who would be able to advise her wisely and unselfishly; one in whom she had implicit confidence. Who so capable and so desirable as Fillmore Flagg? Acting on the impulse of the moment, she wrote the letter directing him to come at once.

To Fillmore Flagg, the summons to Washington

proved as welcome as it was unexpected. He came at the earliest possible moment. The hope of again meeting the noblest, sweetest, and dearest woman in the world for him, his heart's idol; of again being permitted to look long and lovingly into her gloriously beautiful eyes, stirred his emotional nature intensely, and fired his throbbing pulse with the fever of impatient expectancy. The beautiful words of the poet Dennison, in his "Night Ride of a Lover," were ever in his mind and on his lips. Over and over again he murmered:

"Though fleet as an arrow he flies,
Though sundering space swiftly dies,
My heart cries 'Oh haste!
All time is a waste
'Till I drink of her soul at her eyes!'"

The speediest express train seemed a laggard, left far behind in the race of the journey by his swift desire, which kept pace with the telegram announcing his departure from Solaris and the probable time of his arrival in Washington. At length his heart was made glad by a distant glimpse of the dome of the Capitol, which seemed to give him a welcome greeting as it marked his approach to the great city. He found Fern Fenwick's carriage, with Mrs. Bainbridge waiting for him at the depot. Half an hour later he was shown into the library at Fenwick Hall, where in radiant beauty his blushing sweetheart gave him a royal welcome.

As he approached her, with shining eyes and face aglow, soul and body radiant with the grace and adoration of his all-absorbing love, the heroic order of his manly beauty thrilled the heart of Fern Fenwick with its irresistible charm. The kisses claimed by a lover's privilege, she was powerless to deny. Nay!

she did not try to hide the shining light of a great happiness from the adoring eyes of such a noble lover, whose magnetic presence stilled the tumult of her fluttering heart with the ecstatic calm of a measureless content; that unmistakable signature of sanction, that crowning seal of nature's approval which greets the meeting of kindred souls, who, mated in the warp and woof of the web of destiny, in the flashing flight of Cupid's dart, become the harmoniously united halves of a perfect whole.

Ah, thrice happy, thrice blessed, thrice crowned lovers! How swiftly passed those golden hours, as hand in hand, they sat entranced, with soulful eyes in silent communion, dreaming and drifting in the cloud-land of love's harvest-moon, in whose silvery mist they lost all consciousness of the existence in this world of aught else beside themselves!

The next morning after his arrival at Fenwick Hall, Fillmore Flagg having breakfasted with Fern Fenwick and Mrs. Bainbridge, accompanied the former to her work room in the tower. Here, as had been arranged on the previous evening, she gave him a complete account of her work in Washington, since the transition of her father. She also gave the details of her general plan for enlarging the scope of the work to include the women's movement and of directing the combined work in such a way as to become an aid to the work of the model farm.

"My dear Fillmore," said Fern, "How are you impressed by my scheme for carrying out the chosen plans? Can you suggest anything that may be of assistance to me?"

"Your scheme," replied Fillmore Flagg, "is a glorious one which promises to start a revolution in the aristocratic circles of society It impresses me profoundly, as a deep laid plot, cunning and strong, which must accomplish a vast amount of good for the interests of humanity. So deep, so broad and so vast are its possibilities, that a week devoted to study and reflection would but poorly prepare me to understand its significance or perfection as a whole, much less to pronounce judgment upon it. But at this moment, of one thing I feel sure—that the noble purpose which has inspired your skill and genius in the construction of this remarkable plan, which deals so effectively and practically with human life as the result of planetary evolution, will prove a sure guide to success. The plan itself, in all of its details, is already so perfect, in my estimation, as to leave nothing for me to suggest by way of improvement. It is characteristic of you and of your capacity for brilliant work! I am, more than ever before, amazed at this exhibition of your intellectual greatness, which demonstrates your power to think so deeply and plan so wisely. I am very proud of you! I am especially grateful for this opportunity to burn incense as a worshipper at the shrine of your genius! You ask to what extent will the work affect the destiny of woman? I answer, its possibilities in that direction are limitless! They are beyond the power of any living mortal to comprehend! With woman surrounded by such conditions of financial independence, and such harmonious environments as will permit her to devote the best energies of her soul to the perfection of the highest type of motherhood, there will come a solution of the problem of how best to accomplish the perfection of the race. Surely, generations far in the future shall rise up to call you blessed! Dearest, best and noblest of women! Go forward bravely without a fear for the result. Uudoubtedly your plan possesses all the elements of success. With the talisman of your goodness and beauty as the moving force, you cannot fail. Whatever I am capable of doing to assist you, I shall do gladly, with all my heart and strength."

"Thank you, my dear Fillmore," said Fern, "your words of assurance and approval, so beautifully expressed, have appealed potently to all that is good and spiritual in my nature. They have inspired me to better and nobler deeds. They are very grateful to me and I prize them highly.

"Now that you are so much interested, I feel sure you will be able to help me in thinking out some problems which puzzle me. For instance: From among the people I have interested, I wish to select and concentrate the dominant thinkers and workers of both sexes and from all classes, into some kind of a club organization, for the purpose of still further perfecting the efficiency of organized co-operative effort. Question: Shall this society take the form of a club? If so, what name shall I choose for it? In its formation what method shall I use? Can you evolve anything from your inner consciousness in answer to these questions?"

Absorbed in the intensity and earnestness of her questioning spirit, Fern Fenwick left her chair and as her interrogatories came to an end, she stood by side of Fillmore Flagg, looking straight into his eyes with such a penetrating, magnetic glance, that for some moments he was unable to reply. With his beautiful curl-crowned head thrown back to meet and return her entrancing gaze, he breathed but slowly and for the moment seemed rigid as a man of marble; a far-off, dreamy look shone from his half closed eyes. Presently, with a long sigh, speaking very slowly and softly, he said: "Ah! Miss Fenwick, I think I see what you are reaching out for. Your idea is coming to me now quite clearly." Then with returning animation he continued: "Yes, I grasp the idea; it is capital! I believe I can help you. I would suggest the use of the club formation without using the word 'club' in its title. I would call it 'The Twentieth Century Cosmos.' I would choose for its badge of membership a small silver fern leaf, crossed by a large gold key. I would advise that you alone, as the founder and sole director of the club, should have the power to select the members, and to decorate them with the badge of membership. To be in harmony with the century idea, the number of members should be limited to one hundred. All meetings of the club should be held in suitable rooms at Fenwick Hall; these rooms should be known as Cosmos Court. Admittance to each meeting should be gained by the presentation at the door, of an invitation, printed on club paper, bearing the name of the member, giving the date and stating the object of the meeting, all duly attested by your written signature as director.

"The object and purpose of the existence of the club may be stated as follows: That its membership may secure, by the harmonious association of properly qualified minds,—which shall represent the dominant thinkers in all departments of knowledge—a higher, broader conception of the possibilities and purposes of life; as the necessary basis which shall make it possible to acquire a larger store of cosmic wisdom, by the use of systematic methods of co-operative research, study and thought.

"This system of formation for a club would certainly be unique. I believe it will prove to be especially well fitted for the accomplishment of your peculiar work. Does the plan proposed meet your approval by offering satisfactory answers to your questions?"

"Oh! my dear Fillmore," said Fern, "what a darling, clever boy you are, to be sure! Now it is my turn to praise your wisdom and your genius. I think your plan is an excellent one, which will suit the exigencies of my purpose most admirably. Before you return to Solaris we will consider the details more at length. Now let us change the subject.

"In keeping you so long at my work, how selfish and thoughtless I have been! I shall try to make amends! I have planned to make your brief visit as pleasant as possible. To day I must show you over the house and grounds. In the afternoon we shall take a long drive which will give you a glimpse of the beautiful streets, buildings, parks and monuments of our lovely city. Each afternoon these drives are to be repeated, until you are familiar with the great possibilities of this city of destiny, this priceless gift—the perpetual home of the government of the nation—from General George Washington, who is forever enshrined in the hearts of the people as the founder of the republic, the father of his country! When you return to our farm people, I wish you to be able to impress them with the matchless beauty, vastness and importance of the City of Washington, the political center of this unrivaled republic. It is my great desire to have them always think of it and speak of it with love and pride, with feelings of individual proprietary interest, as they realize that they are important factors, as voters and working units of the government, in the great work of shaping its destiny.

"As you are the guest of honor at Fenwick Hall, I am going to do my best to make you, for one week, the happiest man in town! The evenings are to be devoted to the theatre, the opera, and to various society events at Fenwick Hall, arranged for your especial benefit and edification."

"My dear Fern," said Fillmore, "How good and kind you are! To be near you, to hear your voice, to look into your beautiful eyes; is paradise for me! A week so full of happiness, I shall cherish as the one week of a lifetime! As to these society events of which you speak, I shall be jealous of each moment so devoted which shall take you from my side. Pray then, my good angel, do make such moments as short as possible!"

"Rest assured, my knight of the farm, you shall have no cause to complain," said Fern, with a saucy smile as she laid her hand caressingly on his arm. "You are to come with me, prepared to look and listen, while I show you the beauties of my Washington home!"

* * * * * * * * * *

As the "Saint Louis Express" left the Washington station, westward bound, Fillmore Flagg caught a final glimpse of Fern Fenwick, as with characteristic grace and enthusiasm she continued to wave a parting salute with her dainty lace handkerchief, until the train had vanished around the curve. With a sigh he returned to his seat to muse over the events of the week which had passed so sweetly yet so very swiftly for him.

Yes, Fern had kept her pledge up to the last moment. As the guest of honor at Fenwick Hall, she as hostess, in all the graciousness of her bewitching beauty, marked by such charming tenderness, had made him conscious each day that he was indeed the happiest man in town. He now returned to Solaris with renewed courage and enthusiasm, to prepare for the celebration at the farm of the coming arbor-day festival, which Fern had promised to attend. As this celebration was to mark her first visit to Solaris Farm, he wished most ardently to have it prove a great success.

The events of the past week had been a revelation to Fillmore Flagg: a host of new attributes to the noble character of Fern Fenwick had shone forth and dazzled him by their unexpected brilliancy. He began to realize what a wonderful woman she was in this new role, as the queen of the select set in the aristocratic circles of Washington society.

Her strange power to mold the minds of these people; to make them strive for the accomplishment of social and industrial reforms, which meant the redemption of the masses, impressed him most profoundly. By what remarkable process had she, in so short a time, achieved such commanding heights of intellectual and spiritual greatness? Heights, where by operating from the vantage ground of the social and political center of the republic, like some chief marshal on the broad field of human events, she could, by the unseen and irresistible power of hypnotic suggestion, inspire, guide and control the causative and law-making forces which so powerfully affect all social and industrial conditions. Was it possible that spiritual unfoldment alone, could confer such marvelous power? Apparently in response to the intensity of his question, came the reply:

"When a person representing combined physical, intellectual and spiritual unfoldment, is inspired by a noble. unselfish desire to accomplish a great good for all human life, by the use of methods that are in conjunctive harmony with the evolutionary progress of the planet: then such a desire acquires an irresistible force. Naught can prevail against it! In compliance with the demands of a wise cosmic law, it has received the omnistic seal of nature's approval."

The clearness and wisdom of this unexpected reply, appealed strongly to the reason of Fillmore Flagg. Profoundly moved, yet outwardly calm, he perceived at once that the truth of the statement was absolute! In the new light of this remarkable revelation, he wished to carefully examine the claim of the model co-operative farm to the seal of nature's approval. Were the desires, the ideas and the methods in conjunctive harmony with planetary evolution? Apparently they were! That the success of the model farm meant the elevation and future happiness of humanity, was true beyond question. Equally so was the intensity and unselfishness of the desire which had inspired his action and the acts of Fennimore Fenwick and his daughter, Fern. Surely then, the project bore the unmistakable stamp of approval which foretold success! It could not fail! It must succeed! It was irresistible and invincible!

CHAPTER XXII.

A SURPRISE PARTY AND RECEPTION COMBINED.

As the train approached the station at Solaris, Fillmore, in blissful ignorance of coming events, began to prepare himself to leave the coach. In response to a letter from George Gerrish, he had wired from St. Louis the time of his arrival. As he was stepping from the train to the long platform, his hand baggage was seized by trusty hands and quickly disappeared. He noted with amazement the gaily decorated station and the throng of waiting people. Before he had recovered from his surprise, Gertrude Gerrish, evidently striving to assume a very dignified deportment, advanced to meet him. As she gave him a hearty welcome, she said:

"As the leader of the reception committee, representing the membership and children of the Solaris Farm Company, who are gathered here in holiday attire, unanimous in a desire to do honor to you; I greet you! I welcome you back to Solaris Farm!"

Turning quickly, with a wave of her hand, she said: "People of Solaris, three cheers for our General Manager!" At this time, the train having departed, the farm people almost covered the platform with two deep lines, facing a narrow lane in the center, with heads uncovered, prepared and waiting for the signal. The response came instantly in a ringing cheer from six hundred well-trained throats: "Hurrah! Hurrah! Hurrah for Fillmore Flagg! Welcome! Welcome! Welcome back to Solaris Farm!"

Almost before Fillmore was aware of what had really happened, Gertrude Gerrish had taken his arm, as with a mysterious smile she said: "I am now to escort you to the carriage prepared for your reception. We are then to be escorted by the procession to the public square, in front of the hall of education and amusement, where the final ceremonies are to take place. Of course you are surprised! We have planned for that very purpose! So come along now without one word of protest! At the proper moment you are to have as much time as you may desire in which to relieve your mind. For the present you are to keep quiet and obey me—a despotic master of ceremonies whose will is imperative and whose dignity is not to be questioned, even for a moment!"

Fillmore Flagg, now obediently dumb, entered into the spirit of the occasion. He was very much surprised—nay, well-nigh dazed—yet withal delighted, as the happy significance of this unexpected welcome came slowly into his mind. With hat in hand, bowing and smiling, arm in arm with Gertrude Gerrish, he slowly passed between the long lines of happy faces, keeping step with the throbbing measure of the soft sweet music discoursed by the band. At regular intervals, groups of gaily dressed children waved their pretty flags or playfully pelted him with roses. As the twain reached the end of the lines, a novel chariot was waiting: a ladder-wagon of the Solaris fire company, drawn by twenty brawny fire laddies, was equipped with a broad platform, beautifully draped, bearing at each corner a choice selection of fine large potted palms. In the center of this platform was a smaller one, raised still higher; on this was placed the seat of honor, which was covered by a lovely canopy of artistically interwoven ferns and flowers. A broad flight of rough board steps, carpeted and decorated, led up to the lofty seat on this unique chariot. While our hero and the "Master of Ceremonies" were climbing to reach it, the procession quickly formed about the chariot into an elongated hollow square, eight ranks deep; the children with their flags marching in alternating lines of boys and girls, formed the front of the square, while the adults arranged in the same order, formed the sides and the rear. Gilbert Gerrish, with the band of musicians, selected by him from the ranks of the musical club, was placed in front of the square. He was very proud and happy as he flourished his baton and gave the signal for the procession to move forward. In this order they marched gaily along the broad, tree lined avenue which led from the railroad station to the village square. The chariot came to a halt just in front of the hall of education and amusement, with the seat of honor facing eastward toward the center of the public square. The procession quickly reformed into three sides of a square, with the eight ranks facing inward.

For a brief period silence reigned. Then at a signal from Gertrude Gerrish, as Fillmore Flagg arose with uncovered head and stood by her side, the cheers and greetings of welcome were repeated by the ranks with redoubled animation and intensity.

At this juncture, George Gerrish came forward to the front of the raised platform, while Gertrude, turning to Fillmore, said; "The president of the Solaris Farm Company has been chosen by its people to present to you a gift which they have selected, as a tribute of their affection and also of their devotion to you and to Solaris Farm."

"My esteemed friend and co-worker, Fillmore Flagg," said George Gerrish: "As the mouth piece of our people, I am happy to be permitted to join in the active work of this reception. The people of Solaris Farm, moved by one impulse, inspired by sentiments of sincere friendship and enthusiastic loyalty, desire to present for your acceptance, this Solaris album, as a testimonial of their loving admiration; as a token of their absolute confidence in the wisdom of your leadership. This album contains photographs of all the members of the company.

Each picture is endorsed with the signature and with the place and date of birth of the individual. They are arranged and indexed in alphabetical order. Our people were guided to a choice of this gift because they were so profoundly impressed with the importance of the experiment represented by this farm. Because they felt so confident that its assured success would sound the key-note of a general movement for the emancipation and elevation of humanity by the gradual introduction of wiser and better social and industrial methods, which would eventually result in the banishment of poverty and crime.

"Taking this view of the future, we may be pardoned for prophesying that fifty years hence, this album of the pioneers of the movement, will possess a greatly enhanced historical value. We trust, therefore, that this possibility may make our gift more acceptable. I now ask you to receive it in the spirit of love which inspired its donation. In conclusion allow me to assure you that under all circumstances, you can count on the life-long friendship and loyalty of the people whose pictures will greet you, as the years come and go, whenever you may feel inclined to look through the picture laden pages of Solaris Album."

As George Gerrish concluded his speech, a swelling storm of cheers for Fillmore Flagg burst from the ranks of the square. Again and again came the repeated roar of cheers, accompanied by the roll of the drums, and a circling cloud of waving handkerchiefs, hats and flags. Fillmore Flagg, inspired by the enthusiasm and excitement of his cherished people, looked very handsome and heroic as he stood with his manly figure erect, his noble head thrown back, his eyes shining with emotion, the album held firmly in his right hand. Bowing and smiling, he turned gracefully to face the greetings from the ranks of familiar faces, which were swaying with joy and shouting so wildly. Waiting for a few moments, he then raised his left hand, with the open palm outward, as a signal for silence. The tumult was stilled as if by magic.

"People of Solaris!" he said; his clear, strong voice vibrating with emotion: "To you, through your worthy president and your able committee, with a grateful heart, I return my thanks for this most unexpected and charming reception; for this beautiful and appropriate gift, which I prize much more than words can tell. Believe me when I say that I most thoroughly appreciate the noble sentiments which inspired its selection. I am delighted with the happy significance of this demonstration, as a prophecy of the complete success of this experimental farm. This exhibition of your loyalty to me and to Solaris Farm, fills my heart with emotions of grateful joy. You have made me very proud and very happy! I shall never forget the encouragement of your enthusiastic support, which has given me renewed vigor and strength to carry forward the work. I now pledge to you my sacred word of honor that the golden memories of this glorious occasion, and the possession of this precious album, shall henceforth inspire me to still greater efforts for the success of our cherished enterprise, which means so much for us, so much more for humanity.

"I am willing to acknowledge without a moment's hesitation, that your surprise for me was skillfully planned; that its execution was charmingly successful! I wish to return the compliment. I have a surprise in store for you! The present moment is propitious; I will disclose it! I am the bearer of a gift for you—a gift wisely chosen, which is in every way worthy of your admiration and appreciation. A gift of such exceeding value, that I cannot speak of it without becoming eloquent. Gold and silver cannot measure its worth to you! Securely packed in strong cases, which are now lodged in our express office, is a rare collection of books. This collection contains ten complete sets of the best text books for each one of the classified sciences, together with the vocabularies, dictionaries, charts and drawings belonging thereto. Accompanying each set is a miscellaneous collection of the best works written descriptively on that particular science. These books are intensely interesting and very valuable, although they are not classed as text books. Altogether the five hundred volumes form the finest and most comprehensive collection of scientific works I have ever seen. They are the most useful and expensive books published that can be found in the whole range of scientific literature. They contain the knowledge we most need in our enterprise, to enable us as an associated body of people to do better, wiser and more effective co-operative thinking and working.

"To meet and satisfy our needs in this direction, these books were chosen as a gift to our library, by Miss Fern Fenwick, the beautiful and generous patroness of Solaris Farm. She desires me to emphasize her wish that you abstain from any public expression of thanks. In lieu thereof, she prefers to accept the measure of your diligence and enthusiasm in acquiring the stores of knowledge thus offered, as the most appropriate and satisfactory measure of your gratitude to her for the gift.

"To master the contents of these books, is to master the sum of human knowledge in the various departments of science. With this mastery there will come to us the largest understanding, and the clearest obtainable conception of our relations toward each other, and to the universe around us. Thus enlightened, we may discover that ignorance is a sin; that as responsible entities in the great pulsing sea of cosmic life, with more or less power to help or hinder the purpose and perfect unfoldment of all life —we cannot afford to be selfish, sinful or cruel in our actions toward each other, or toward any other form of cosmic life. Having once acquired these convictions, with this most important fund of information, we possess the key which will unlock the mystery of the action and reaction of the potent and unseen forces of nature, which affect us as individuals, as they do the earth, air and water, the elements so necessary to our existence. The restless, never-satisfied, questioning spirit, born with every human soul, is the expression of a divine purpose! To gratify this insatiable desire for more knowledge, is to comply with the demands of a wise cosmic law. By so doing, we enter into the enjoyment of a never-failing source of perpetual delight. We are crowned with a happiness of the purest type!

"In viewing this vast field of knowledge, spread so invitingly before us; in anticipating the joy we may glean therefrom; we catch a glimpse of the exceeding richness of the boon of immortality, which, as a spiritual heritage, is waiting for us. We begin slowly to understand ourselves as the repositories of infinite possibilities!—as cosmic units of the larger Cosmos—as a perfect microcosm of the macrocosm! With feelings of awe-inspiring adoration, we reflect that we may know ourselves as individuals, only as the extent of our knowledge of the universe around us is increased. Responding to the law of action and reaction, the more we reflect, the greater becomes our desire to know more of ourselves. Always more! Ever more! Never quite satisfied! Fortunately, the immortality of the wisdom loving human soul embraces all time, and all eternity! Therefore, through the law of eternal progression, we may naturally and rightfully aspire to the acquirement of all possible knowledge. In cultivating these aspirations, we may rest assured that we shall constantly gain new conceptions and new meanings for the word 'Heaven.'

"In conclusion, my friends and co-workers, my brothers and sisters, let us congratulate ourselves as the fortunate recipients of this priceless gift: let us endeavor to show our appreciation by a speedy mastery of the contents of these valuable books. Let us approach the work, full of joyful anticipation and enthusiasm, with the proud consciousness that we are invited guests to a great feast of learning. Let us strive in every way to make study thoroughly enjoyable. Let us make it one long holiday in honor of the Goddess of Wisdom! One grand harvest-home of our gathering of the golden fruit from the tree of knowledge. Let us be as earnest as we are enthusiastic—let us be thorough, and withal methodical and systematic.

"The ten sets of text-books, suggest the formation of the membership of the company into that number of scientific clubs; which I recommend. This division would give fifty adults as the average membership of each club. We have at least ten available rooms large enough to accommodate clubs of that size. Each club should begin with the primary text-book, which should be read, discussed, analyzed and re-read until clearly understood by the entire class. The club to proceed in the same order with the next of the series, until all are thoroughly mastered. I will volunteer to join the club to which is assigned that scientific study which may prove the most difficult, least inviting and most unpopular. By the force of a united purpose, working co-operatively together, we shall soon develop a capacity for severe mental labor, which will make the mastery of the remainder of the course a constant source of pleasure. What we need in the way of equipment, chemicals, instruments, etc., can be easily and quickly secured.

"George and Gertrude Gerrish will have an advisory superintendence over the work of all the clubs. Years of experience in teaching have prepared them to quickly untangle the mixed quantities or conditions that may confront us, and thus skillfully turn our difficulties into delights.

"With this general plan for conducting our literary festival, I will leave the subject with you for consideration at the proper time.

"I feel conscious that under the circumstances, I owe you an apology for having so trespassed upon your patience and good nature, by the length of my remarks. Therefore I desire to acknowledge my thrice doubled appreciation of your manifest interest, attention and sympathy, which have both flattered and encouraged me greatly.

"I will now close by thanking you, through your worthy officers, for this cordial and beautiful reception; also for the opportunity to address you on a subject in which I am so deeply interested."

CHAPTER XXIII.

FORMATION OF POPULAR SCIENCE CLUBS.

As the days passed after the reception, the new books were unpacked by Fillmore Flagg, assisted by George Gerrish. As soon as possible they were arranged and placed on appropriate shelves in each one of the ten rooms prepared for them. Large steel engravings in plain oak frames, of all the authors, together with the maps and charts, all neatly glazed and mounted, adorned the walls of the particular room to which they belonged, adding greatly to the attractiveness of the general collection. As the work progressed, the keen interest displayed by all members of the farm company seemed to increase. They could talk of nothing else; they were eagerly and almost impatiently waiting for the announcement of the formation of the clubs. Accordingly therefore, as soon as the rooms were ready, a complete schedule of the books in each series was made; these schedules being numbered from one to ten, to indicate the series to which they belonged. They were printed and distributed among the members of the company, with a request that one week later, each member should return two of the numbered schedules marked as first and second choice of the studies they desired to take up. By this method of voluntary selection, the clubs were quickly and easily formed, without friction or embarassment. Well stimulated by an ever increasing fund of interest and enthusiastic ambition, the club members, impressed with the wisdom of Fillmore Flagg's advice, promptly took up the class work of the study chosen, eager to secure a generous share of the educational benefits to be dispensed at the board of this great literary feast, to which they had been so kindly invited as especially selected guests. With some misgivings as to the final result, Fillmore Flagg carefully watched the preliminary club work while yet in its organic stage. He had been somewhat doubtful of the ability of the average club member, who was not a trained student, to acquire a sufficient interest in such abstract subjects, with which to develop the mental force so necessary in order to digest and finally master them. However, much to his surprise and delight, at the very threshold of the work, the display of energy, ability and mental acuteness on the part of the entire club membership, dispelled the last remaining doubt from his mind; he was convinced of the practicability and final success of the course.

In carefully analyzing the subject, he perceived that they were quickened by the momentum of a united co-operative effort; also that they were—perhaps subconsciously—pushed forward by a great number of new ideas concerning the desirability of at once acquiring a larger store of scientific lore, as a necessary and more complete equipment for the practical duties of the battle of life. Dominant and central among these ideas, was the one which so temptingly promised an increased knowledge of themselves as individuals, by the mastery of the broad and hitherto unexplored field of explanatory science; which might lead to a better solution of the mystery of environmental conditions. Finally, they were no doubt inspired strongly by a firm conviction that, once armed with a thorough scientific education, they would possess an additional power to aid in making Solaris Farm a speedier and more pronounced success.

Fillmore Flagg accepted this demonstration of the combined ability of the farm people to conquer the most difficult problems of science, without the advantage of previous training, as an added proof that the ideas and methods of the model farm were most assuredly in conjunctive harmony with planetary evolution; therefore with the great force of combined co-operative mental effort to push it forward, still more surprising results might reasonably be expected, when these efforts were more wisely and skillfully directed along lines indicated by nature as lines of the least possible resistance. A realization of these expectations would seem to suggest that the key to future success in all educational work lies in discovering systems, methods, associations and surroundings for the students, which are nearest in conjunctive harmony with natural evolution, consequently along a pathway presenting the fewest possible obstacles.

CHAPTER XXIV.

A TWENTIETH CENTURY LOVE LETTER.

"ALL THE world loves a lover!" is a trite but beautiful saying, which touch a responsive chord in the great heart of humanity! We cannot remain indifferent to the magnetic effect of the strong tide of his eloquent and impetuous wooing. Nor can we withhold a sympathetic desire to aid him in reaching the goal of success—to win the precious prize. Quite as naturally, we are intensely and delightfully interested in the birth, the unfoldment, and the blossoming of every individual entity in the great ocean of cosmic life. Instinctively we recognize that love is life. One could not exist without the other. Old and young alike understand the potency of the spell which binds the lover; which holds him for unconscious periods of time, absorbed in dreamy contemplation of his ecstatic devotion to the heroic virtues, graces, accomplishments and attributes of the charming woman, whom his heart has chosen to represent all things in the universe which have meaning and worth for him. Through this adorable woman, the crowned and glorified object of his all-absorbing love, he can best respond to the rythmic throbbing of all cosmic life. In this superior state of beautiful transfiguration, he forgets self, and lives for long happy months in the rare upper strata of real unselfishness. Under the powerful influence of pure love, the highest and holiest emotion which stirs, controls and makes better the life of every mortal; lost in the blissful alembic of this great chemical change, the lover recognizes himself in every demonstration of universal life around him. Ho also becomes aware, from some inner consciousness, of the extent to which the emotional nature controls and molds the individual; that among the anabolic emotions, love is the queen of the emotional empire; that the touch of her magical scepter is so potent and penetrating as to render the individual receptive and responsive to all of the ennobling, purifying, progressive and exalting elements of the universe: but, on the other hand, what is still more marvelous: that the same touch renders the individual negative to the inflowing currents from all of the baser elements. With this awareness comes the conviction that the Empire of Love is boundless and limitless; that it permeates and glorifies the vast ocean of infinity! On the strong, swift tide of this shoreless ocean, the lover floats, secure, serene and confident, on his voyage toward destiny's most distant port.

The following letter from Fillmore Flagg to Fern Fenwick, will serve in some measure to illustrate the power of love to change, expand, energize and spiritualize the entire character of the lover: to purify and strengthen the moral disposition of our hero, to eliminate from it all tendency to selfishness; to endow him with a broader wisdom, with higher and nobler aspirations of life; to fit him more perfectly to carry forward his great work for humanity at Solaris Farm.

* * * * * * * * * *

"My Darling Fern: Noblest, purest and most beautiful of women! Like the rose to the sunlight, like the needle to the pole, my heart turns in adoration to you. My own true love! My peerless one! My guiding star in love's azure sky! My soul swells and sings with its full tide of joy, as willing fingers attempt to put in words the thoughts born of my great love for you. What miracle have you wrought for me, my precious one, that I am so happy? The earth, the sky, the verdant woods, the grand mountains, the green meadows, the shady nooks, the babbling brooks;—all thrill my innermost being with a thousand new charms! The bees, the birds, the flowers and trees as they bend or sigh to the passing breeze; the solemn stillness of majestic night; the deep blue sea, overarched by nature's matchless crown of diamonds, a countless multitude of brilliant stars, in the silvery moonlight of love—how eloquent their song! All things in nature speak to me; they bless you for loving me! In the halo of that blessing, as I think of you, I am transfigured by a newly-born ecstacy! To breathe, to exist, is to realize the superlative degree of my exquisite happiness! Hidden away from the clouds and storms of life, by the golden mist which veils the measureless sea of love, infinite love, I sail serene and confident upon its heaving tide. Gently rocked by the lapping lullaby of the rythmical waves of paradise, I fearlessly float. I care not for time nor tide, nor distant port of a future destiny! Entranced by the music of love's beautiful sea, I dream love's dream alone with myself, the outer world shut away—swallowed up by the overwhelming tide of my sweet and blissful contentment.

"From such hours of exaltation, I am sometimes rudely awakened by a monster reflex wave of self-examination. Ah, dear heart! It is then that I ask of my soul: What am I? What have I done? What sweet guardian spirit guides my life, that I should be made so exceedingly happy by the priceless love of such a beautiful woman? Am I worthy of such a blessing? Can I properly appreciate the great good fortune of being fondly and truly loved by such a peerless woman, who is so dear to me, so noble, so good, so true; so pure, so bright, so beautiful; so truly wise, so eloquent; in every way so well fitted by birth, wealth, and education to reign as queen in the most brilliant and most exclusive circles of the social world; even in the grandly beautiful city of Washington, where the princes and potentates of the earth, lords of other lands, of wealth and fashion of high degree, vie with each other and with the republic's most honored statesmen, for one smile, one look of recognition from this marvelous woman, who is everywhere recognized as the dominant center of attraction? Oh, the wonder of it! This is she who holds the key to my heart!

"Ah, my adored one! As this picture of your life fills my mind, I wonder what would happen to me under such circumstances, with any other woman in your place. I know I should be both furiously jealous and foolishly despondent: but with you, the very apotheosis of truth and honesty!—Impossible! It could not be: so base a thought would perish with the thinking! I know you are as true as steel. The pure soul which shines from your eyes has spoken to mine. I am content; I fear not; I know that the compass of your love is constancy.

"Oh! my darling! Chosen one of my soul! How great is the mystery of love! How priceless the blessing it brings to the lover! How brilliant the constellation, how spiritualizing the multitude of new thoughts to which it gives birth! How I pity those who have not been touched and quickened by the life-giving power of love! How sad and desolate is the pathway of the soul so unfortunate as to be shut away from the sunshine of love! Better, far better, to die of love! To die of love is to live by it! It is to have discovered the great deeps of the infinite: for love itself is a revelation of the infinite! The aspiration of love is the inspiration of paradise. Who can understand the significance, or the great mystery of immortality, or the fulness of the promise of eternal happiness to be gained by a life of endless progression, without first having lived a life of love? The smile of love is the rainbow of life! Every tender emotion of love is a prayer, pure and potent, for a higher life.

"The truth of these things, my sweet heart, I realize more fully each day. I feel and know that every link in the chain of eternal existence, is a link of love! My love for you has been for me a spiritual blessing indeed! It has opened the eyes of my soul, so that I may perceive the significance of the miracle of love, which must precede the miracle of birth, as the necessary beginning of the unfoldment of the individual up to his highest estate—the repository of infinite possibilities. Love, then, my dear one, is the highest and holiest attribute of the human soul: that inspiring, controlling force, which wings the soul to such sublime spiritual heights, as are far above and beyond the storms of common passions, and the evil influences of the baser emotions.

"Ah! sweetheart of mine! How much do I owe to the uplifting power of love! I question and wonder! When its divine radiance shines upon me, through the glory of your beautiful eyes, I am led up the steep acclivities of the mountain of wisdom by a new pathway. I perceive that as the oracle of life, love is the potency which crowns woman with that entrancing aura of soft, sweet, melting force, which for ages has proclaimed her the greatest and most fascinating mystery of the universe! I also perceive that, responding to the stimulant of this potential aura, I am thrilled, spiritualized, energized, encouraged and more perfectly fitted to perform whatever difficult or heroic work the needs of our farm people may demand. Fortunate for me was the day when Fennimore Fenwick left you heir to his plans for redeeming the lives of these people! Fortunate indeed, was the time when I was chosen by you to discover, select and institute Solaris Farm, with the broad humanitarian work which its success represents. Each memory of this farm; of my every thought, plan or deed for its improvement: of its people; of their lives, health, and happiness; of their sublime confidence in me, of the prompt obedience they so cheerfully render to my slightest command; of the peculiar pride expressed by the appreciation of their importance as working units of the farm, all united, harmoniously blended, in one perfected co operative mass;—is a memory made more delightfully permanent by the wonderful light of your love!

"Never before have I been so busy or so blessed! Every emotion of pride, enthusiasm, ambition, joy or love, which stirs the hearts and quickens the pulse of these people, who are working with me for one object so faithfully, so earnestly; through the magnetic halo of your love, is reflected upon me with redoubled intensity. In the strong current of this electrical stream of power, I am quickened, strengthened and prepared to do better thinking and more effective work for the perfect development of the farm.

"At this point, dear Fern, I must mention an item of farm news, in which I am sure you will be greatly interested. We have arranged to have our arbor-day celebration, or tree planting festival, on the 10th day of the month of March in each year, as the season, in this climate most suitable for the work. For some months past, for the purpose of exciting in the minds of our people a keener interest, I have been giving a course of lectures on the general subject of forestry. These lectures have proved so attractive, that as a result, they have been exceptionally well attended by both old and young. The amount of interest displayed by my hearers, is a continual source of surprise and delight to me. Early in the course, this extraordinary interest culminated in such a perfect shower of questions in regard to the details of the subject, that I was obliged to refer my questioners to the various books written on the subject, as most completely and satisfactorily answering the multitude of their queries. As a consequence, the botany club has had a great boom. While every book in the library on forestry, or the care and culture of plants and trees, including those in a full series of annual reports from the Department of Agriculture, is in constant use. You would be delighted, my dearest, could you note the readiness of even the children to grasp the idea, to understand the immensity of the benefits which may be conferred on future generations by our systematically directed efforts in tree planting here on this farm. Both young and old alike, are quick to appreciate the important fact that while we are enjoying a holiday, to which we may look forward each year with increasing delight; we are at the same time furnishing the world with an object lesson as to the practicability and great value of the good work which may be accomplished by all classes of agricultural people, in the general observance of such a festival.

"The announcement of the good news that you are to visit the farm in time to attend our first arbor day celebration, on the tenth of next month, has made our people very happy. They are simply wild with delight at the prospect of seeing you so soon: of having an opportunity to thank you in person for the many favors you have so generously bestowed upon them. Hitherto they have admired and adored the beautiful and generous young patroness of Solaris Farm, through the medium of a life-size crayon portrait, made some months ago, from one of your recent photographs. Since then, this lovely shadow of the idol of my heart, adorned by a suitable frame, has occupied the post of honor, as the only picture on the walls of the library The advent of such a charming picture, at once converted the library into the throne room of the village, where gathered daily, admiring throngs of our people to feast their eyes in silent worship at the shrine of this life-like shadow of your lovely face. In thus exposing this picture, so dear, so sacred to me, to the earnest and respectful admiration of our people without your knowledge or consent; I trust, Dear Heart, that I may not have outraged your sense of propriety in the slightest degree. It occurred to me that it would be just and right, also most fitting and proper that, as the patroness of the farm, your portrait should appear in the place it now occupies; that it would be the most appropriate method of linking your individuality, in the minds of our people, with the peculiar work and destiny of the farm. If you consider my action from this point of view, I am sure you will approve. Like some good fairy, the silent charm of your portrait has each day, each hour, wrought its perfect work in my life and in the lives of our people. It has proved a constant source of delight! An added talisman to insure the final success of our enterprise!

"Ah, my good angel! my Princess Charming! At last comes the crowning thought which completes my wreath of happiness! It comes to me daily, again and again! It is this, Dear Heart; that every step toward the final and complete success of Solaris Farm, is an added link in the chain of a shining destiny which shall bind our lives more firmly together, until at last this beautiful chain of love shall have become proof against the dissolving power of the passing ages of an Eternity!

"In conclusion, sweetheart, may a bright band of

faithful guardian spirits, ever watchful, ever near, guide and guard you, the crowning treasure of my life, is the earnest prayer of

"Your devoted, loving and loyal,

"FILLMORE FLAGG."

CHAPTER XXV.

THE REPLY.

"MY DARLING FILLMORE: Words fail to express the happy effect of the pleasing emotions that arise as I muse and dream, build castles in the air and indulge myself, again and again, in the luxury of reading line by line, the glowing tributes of love in your marvelous letter. I am electrified by its wonderful logic, rythm and melody. Ah, my chosen one! So manly; so noble; so true! The witchery of your eloquence is a conquering force, that Cupid with his bow might well be proud of! My heart rejoices under the influence of its magical spell! I am so happy and so proud of you! The great deeps of my emotional nature have responded to the poetical sublimity of your charmingly expressed sentiments. They thrill my soul like the dawn of some glorious summer day; like the exquisite perfume of a sweet flower; like that sublimely sweet surprise which steals over the senses, while a fleecy veil of silvery mist, responding to the power of the advancing king of day, slowly rises and discloses the shoreless grandeur of that tidal mystery, the majestic, restless, billowy bosom of Old Ocean; like some grand symphony of masterful music, penetrating and resonant, with that mysterious potency which awakens every echo of the soul's musical possibilities! Yet, sweetheart, every word is charged with your personal magnetism; is stamped with your individuality; freighted with the wealth of your spiritual and intellectual development. In every line, sentence and paragraph, I recognize you as my ideal of a lover, the dearest and most noble of men!

"In my retrospective moods, the cloud of memories, born of the incidents which have marked our past acquaintence, form a telescopic vista. Through this vista, examined in the crucible of much correspondence, the intimate association and the mutual friendship of many months duration, I perceive that I have discovered and have learned to appreciate the sterling worth of your character. Through this avenue I become conscious that you represent to me the superior nobility of true American genius; the highest and grandest type of manhood! Idealized as my hero, I place you in the front rank of America's dominant thinkers; a peer among peers, both potential and progressive—yet withal so modest, so free from dogmatism.

"I seem to feel intuitively that you are standing at the very beginning of a new cycle in the history of our planet: a cycle in which symmetry of mind and power of brain, fix the standard by which nature selects the leaders she deems most worthy of ruling the destinies of her people. I feel that you have been measured by such a standard, and chosen as the instrument for the accomplishment of a special work of the utmost importance!

"This bit of hero-worship on my part is due, no doubt, to the intensity of my devotion to our Republic; to the earnestness of my convictions in regard to its manifest destiny as a saving power—an uplifting force—among the nations of the earth. These growing convictions are emphasized by the keener perceptions of my spiritual nature, which declare that this almost resistless force which dominates our Republic, that may be likened to the world's storage battery, is due to the progressive power gained by the universal enlightenment of the American people as a mass. This important thought seems to emphasize the wisdom and the importance of universal education.

"I must now refer to a matter mentioned in your letter, in which I am particularly interested. In declining to become jealous of the bevy of titled lords, who pay fawning court to my wealth and social position, here in Washington, you do yourself justice; while at the same time, you pay me the compliment of a lifetime! When compared with you, how puny and feeble are the princes and titled lords, made by kings and courts, in lands where selfishness reigns supreme at the expense of millions of unfortunate subjects! An impecunious host of these fortune-ihunting lords swarm in the society of our large cit es. With faded titles of doubtful value, as their only stock in trade, they fittingly represent the decaying nobility of passing monarchies. They are looking for victims! They become the highly honored guests of selfish, title-crazy, match-making mothers! Oh the pity of it! Oh the shame of it! How American girls, who are born to wealth, with all of the advantages which wealth may command, including the best education possible in this land of progressive liberty; who should love devotedly the vital principles of our democracy;—can be so dazzled by the false glitter of a title, that they deliberately choose to mate themselves (and their riches,) with such sorry specimens of lordliness; such brainless, nerveless bundles of selfishness, is something too monstrous for my comprehension!

"Are these girls really Americans at heart? Do they represent the women of our land? Can they understand or appreciate the privilege as a birthright, of proudly taking an honored part in the coming motherhood of this great and progressive land of republican liberty; a republic which to day stands as the hope of the world? Is it possible that they can knowingly wish to become mothers of a feeble race of puny children—children who are cruelly bereft of moral, physical and intellectual vigor by the tainted heritage which, like some avenging nemesis, through the action of an inexorable law, surely follows the unfortunate offspring of lordling fathers, who are born as the very dregs from twenty generations of the vice and depravity of kingly courts?

"My dear Fillmore, to these interrogatories I answer, No! A thousand times No! Ignorance! A shameful ignorance of the true object and purpose of human life, on the part of these misguided girls, is their only sin. They are well-nigh hopelessly ignorant of the significance, or even the existence, of the great basic truths of evolutionary life. They know not that each age in the series of evolution grows out of the preceding one; that each in its order is the parent of the next; that the same is true of each generation of people. In the midnight darkness of their ignorance, they are incapable of knowing that virtue inherently possesses the germ of perpetuity. They can neither understand nor heed the warning cry of history, which proves that crime and depravity have in themselves the seeds of natural death. They have never read history's tragic story of the total extinction of the royal houses of Capet, Valois, Tudor, Stuart and Bourbon;—a story which demonstrates so conclusively the avenging results that follow the crimes of royal fathers.

"To redeem these girls from such dense ignorance; to rescue them from the thralldom of such a fashionable sin, which threatens to become a fad; to open their eyes to the horrible consequences which follow such mesalliances, is a work so important as to demand the immediate attention and united effort of a host of America's patriot mothers.

"Pardon me, dear Fillmore, for devoting so much space in my letter to this particular topic. I feel sure you will kindly excuse any excess of fervor which may have marked the expression of my indignation. Because you so well understand the intensity of my devotion to the broadly progressive principles of our matchless republic, you may, consequently, guess the full measure of my scorn for this foolish, title-hunting class of creatures who, like silly moths, blindly sacrifice themselves in folly's funereal flame. The bare idea of marriage to gain a foreign title has always been exceedingly repugnant to me. With passing years, I am each day more thankful that since my early childhood there has been buried deep in my heart, a determination that when the time came for me to select a husband, the only title of the one chosen should be the stamp of honor which marked him as a true type of an American citizen—a real American genius; a truly noble soul, perfectly and beautifully expressed by a harmonious combination of physical and intellectual development!

"Fortunate the day for me when that lucky advertisement brought you to my side, as a trusty, capable co-worker, whom I have learned to respect, to admire and to love. My dreams have been realized. I have found my ideal. You may fearlessly trust in the absolute truth of your assertion that 'the compass of my love is constancy!'

"Now my hero! My ideal of a gallant Knight of Most Excellent Agriculture, whose nodding plumes, of tassels of corn, artistically interwoven with splendid pompons of waving wheat, barley, oats and rye have so dazzled my eyes and charmed my heart; having chanted my song of love, I hasten to assure you that your last report concerning the administration of the affairs of the farm, has pleased me greatly. I think the progress achieved in so short a time, is truly marvelous! Only my Fillmore could have accomplished so much! I am full of curiosity about the details. When I come, you must be prepared to answer a host of questions; to go with me on many excursions of discovery before I shall have completed my tour of agricultural investigation.

"I approve of the disposition you have made of my portrait. Of course my personal pride is gratified by the sincere admiration and praise it has excited. I am happy in the knowledge that it has proved so efficacious as a talisman of good fortune for the farm. I think I understand your reasons for the feeling that my individuality should be in some way directly interwoven with the destiny of the farm.

"Reasoning from the peculiar environments which so affect our lives, I realize more fully each day that my personal interest in every step toward its final success, must necessarily be quite equal to your own.

"I am delighted with the idea of being present at your first Arbor day celebration. I hope there is to be in the order of exercises an oration which you are to deliver. If so, I know you will not disappoint me! I am prepared to prophesy that you will do yourself justice, do credit to Solaris and at the same time you will cover the subject with a halo of glory. Such a result seems assured when I consider the extraordinary interest which was aroused by your lectures on forestry. This signal conquest of your eloquence has gratified my pride very much. I am strongly impressed with the vast importance of this tree-planting school, which you are about to institute at Solaris. The success which you have won in the preliminary work is so promising, that I am sure you have undertaken a task which is worthy of your genius. In my judgment, you have already demonstrated your ability to accomplish many wonderful things. Great opportunities are before you. By the force of your logic, by the earnestness of your eloquence, you will be able to instill and to permanently fix in the minds of our people—both parents and children—the true progressive principles of American citizenship. You will thus enable them to perceive the serious import of the responsibilities which, like a mantle of power, descends upon them, as the representative working units of this great republic. You can so inspire them that they will be eager and proud to take up with honor the burden of these responsibilities. You can so change and elevate the lives of these people and a multitude of others, that first they shall become masters of themselves; later, masters of the republic; through the controlling force, the imperial dominancy of scientifically developed, symmetrical minds; whose intellectual, ethical, inspirational, logical and constructive power, combined as an elevating agency, shall raise the republic of the future to still more com manding heights. To accomplish these things, is the glorious beginning of a great career! In visions of your life work, it comes to me that this preparatory work on the farm is but the introduction to a more important mission, in the vastly wider field of a near future. In this coming work we shall stand side by side. Hand in hand, with hearts united by the bonds of a supreme love, we shall go forth armed with the power to overcome and to conquer the great hosts of ignorance and selfishness which so hinder the world's progress.

"Really, my true love, although this letter is so long, I cannot close it without again expressing my appreciation of your soul-satisfying letter; so laden with the fragrance, the benediction of your love; so potent with the charm of happiness for me. To its benign influence my heart responds by the awakening of the highest and best emotions of my spiritual nature. Written in clear, plain English, it appeals to me as a letter of such sterling intelligence as only my ideal of a lover could write. How different it is from the soft, sweet nonsense of fashionable fops; the effusive gush of poetical dudes.

"Now, I must say to you Good bye, my sweetheart! Remember that waking or dreaming, I love you truly. Only you, so dear to me—you, so generous, so noble, so good. Bright are the links of love's golden chain which time cannot sever. Constancy, our love shall bless, now and forever. May the sweet guardian spirits who guide your footsteps, keep you safely until we meet again, is the ever-present thought which is inspired by love's whisper in the heart of your devoted,

FERN FENWICK."

CHAPTER XXVI.

FERN FENWICK ARRIVES AT SOLARIS.

Fern Fenwick, accompanied by Mrs. Bainbridge, arrived at Solaris on the afternoon of the third day previous to the tree-planting festival. When the train reached the station, they were met by Fillmore Flagg accompanied by George and Gertrude Gerrish, the committee representing the farm company. With this escort to the village, they were soon installed in a handsome suite of rooms, beautifully decorated and furnished for their reception.

After a late luncheon, Fern Fenwick gave a private interview to Fillmore Flagg. During this interview, which lasted more than two hours, matters both of business and of love were discussed: love, however, claimed the lion's share of the time. Very soon, by mutual consent, the major part of the business was postponed until after the tour of the farm, planned for the following day, had been completed. Then with a sigh of relief, they resigned themselves to the sway of that potent charm of blending magnetic and spiritual auras, which so swiftly transports re united lovers to a paradise of their own.

In accordance with previous plans, the next day was spent by the visitors in driving about the farm. The first motor carriage was occupied by Mrs. Bainbridge accompanied by George and Gertrude Gerrish, Fillmore Flagg and Fern Fenwick following in another. Pursuing a carefully arranged program, all points of interest were visited; the barns and stables, herds and flocks, the meadows, the cotton and grain fields, poultry yards, dairy, apiary, gardens, mills, store-houses, packing-houses, factory buildings, the brick works and pottery, the clay-beds, stone-quarries, coal and other mines.

This tour of inspection, which occupied nearly the whole day, proved very interesting to Fern Fenwick. With her note-book in hand, and her keen eyes on the alert to catch every salient point, she kept our hero busy answering a host of questions. It was a long, happy day for him! To sit so near her, to look into her smiling eyes, to listen to the musical tones of her voice, to answer her swiftly spoken questions, to respond to the pressure of her gloved hand upon his arm as she directed his attention to some particular object; all seemed to him such a delicious bit of experience, that he almost wished it might go on forever!

In the evening the reception given in honor of the Patroness of the farm, was held in the large hall of education and amusement. In this hall, which was handsomely decorated for the event, the people of Solaris were assembled. They were a unit in eagerness to give expression to demonstrations of delight when, for the first time, they were permitted to greet the one they wished to honor: a woman whose name they reverenced as the title of the noblest guest they could ever hope to entertain. George and Gertrude Gerrish, with Mrs. Bainbridge, were already seated on the stage, when Fillmore Flagg appeared, escorting Fern Fenwick from the waiting room. Moved by one dominant impulse, the entire audience arose to receive her. The repeated cheers of welcome were intensified by the accompaniment of a fleecy cloud of waving handkerchiefs.

Our heroine was well worthy the ovation: richly and artistically gowned, she was a perfect picture of loveliness! Her cheeks flushed with the excitement of such an unexpected demonstration, her beautiful eyes flashing with the inspiration of her wonderful enthusiasm, her perfect figure proudly erect with the grace and dignity of an all-conquering magnetic presence, she captured the hearts of the people even before she had opened her lovely lips to address them.

Warned by a gesture from Fillmore, the cheering ceased and the audience became seated. He then introduced Fern Fenwick by a neat little speech which provoked another storm of applause more demonstrative than the first.

When order was again restored, at a signal from George Gerrish the double quartet of mixed voices, which had been selected from the singers of the musical club, came forward and, in a style which reflected much credit on the club, gave a song of welcome composed for this particular reception, and entitled; "She comes, she comes, she comes to us; our wise and lovely patroness." This song, which created a real sensation, was followed by an eloquent address of welcome delivered by George Gerrish in his official capacity, as president of the company. His remarks were seconded and emphasized most vigorously by long continued demonstrations of approval from the assembled members.

In response, Fern Fenwick replied at some length in her most charming manner. Turning to George Gerrish, she said:

"To you, the president, and through you, to the officers, members and children of the company here assembled, I offer my sincere thanks for the honor conferred, and for the pleasure given to me by this delightful reception. The sentiments of kindly greeting, of keen appreciation, of admiring approval, so beautifully expressed in your address of welcome, have touched me deeply. I am so profoundly moved, that my heart overflows with grateful emotions! Equally charming, and even more gracious to me were the words and music of the song which your sweet singers have rendered so artistically. These testimonials have so wonderfully impressed me that I can not forget them! As the years come and go, I shall cherish the bright memories of this eventful evening, as added jewels with which to mark and adorn the shining links, interwoven with the chain of my experience in life. These memories shall also serve to strengthen my already intense interest in this most extraordinary farm. A farm with such a wide range of improvements; with such an imposing collection of large well constructed buildings; with so many profitable allied industries in the full tide of successful operation; with a general equipment so magnificent, that at every turn I am astonished and delighted. I now understand why and how you have succeeded in transforming the hated drudgery of farm labor into such a pleasant, desirable occupation.

"Since the beginning of the enterprise, my interest in the work has been constantly stimulated by the detailed accounts contained in the full weekly reports furnished by your general manager. These reports from time to time, I have studied carefully. Therefore I came here expecting much. However, after my tour of inspection, I hasten to assure you, that I was not all prepared to find such an ideal farm, already in successful operation! A farm with proportions so generous, an equipment so complete, and a future so promising; that when I pause to contemplate the magical changes wrought upon it in the brief space of thirty months, I am filled with admiration for its wonder-working, epoch-making people! I consider it a coveted honor to be known as the patroness of such a grand institution. People of Solaris, I am happy to be thus identified with you. I am proud of you and your work! A work which shall yet cause millions to rejoice! You cannot guess; no one can even estimate, the exceeding value of this work as a shining example of what properly organized labor can accomplish. You have succeeded far beyond my expectations! Do not waver or turn aside for one moment! Go forward bravely; be strong and steadfast; be encouraged with the assurance that all times, I am ready and willing to assist you in every possible way! Success with her golden crown waits to reward you! All the world is watching and waiting for the victory, which you have already won. Therefore, in the name of humanity, I am justified here and now, in thanking you for this superb lesson in unselfish co-operation. This lesson in self evolution, which you have given to the world, is a result on your part as individuals, of a wise exercise of mutual trust and confidence in each other; reinforced by the combined industry, zeal, persistence and skill displayed in your noble efforts. By such efforts you have made the name of Solaris justly famous throughout the length and breadth of this Republic!

"In conclusion, Mr. Chairman, and friends, allow me to again express my thanks for your greetings of welcome, and for every demonstration of loving appreciation which you have so generously showered upon me."

While the hall still rang with the plaudits of a delighted people; before Fern Fenwick could move towards her seat, George and Gertrude Gerrish and Fillmore Flagg all hastened to her side, to offer congratulations on the eloquence and excellence of her impromptu address. To the observer, it was plainly evident that the effect of such a stirring speech on the assembled co-operators was unusually impressive. They seemed to be inspired with a deeper reverence and a more perfect loyalty of devotion for this remarkable woman, who had so charmed them by the power of her eloquence. Swayed by the intensity of this deep feeling which could not well express itself in noisy cheering; they eagerly pressed forward in a quiet orderly way toward the stage, where George Gerrish was waiting to introduce them individually to our heroine, the patroness of the farm. Smiling graciously as they approached and were presented, she took each one by the hand in such an earnest cordial manner, that all feelings of shyness or embarassment were quickly banished. After the exchange of a few words of pleasant greeting, they quietly returned to their seats. As the reception progressed, many of the members improved the brief moments in expressing their grateful appreciation, for the words of praise which she had so enthusiastically bestowed upon them, in a speech they could never forget.

When all were again seated, George Gerrish announced that the program for the evening would close with three short selections, to be given by volunteer members from the ranks of the musical and dramatic clubs. With this part of the entertainment finished, before the people could be dismissed, Fern Fenwick arose to bid them good night, and to thank them for such a charming reception, which she pronounced "simply delightful!"

CHAPTER XXVII.

THE FESTIVAL.

FORTUNATELY for the tree-planters, the day of the celebration at Solaris, proved exceptionally fine! No one could resist the exhilarating tonic of such a perfect day! A day made more glorious by a cloudless expanse of blue sky, a flood of golden sunlight, and breezes, soft as the balmy breath of gentle spring could make them!

The tools and the potted trees, each labeled with the name of the planter, were hauled in wagons from the nursery to the site of the future forest, where the ground had already been prepared to receive them.

At nine o'clock in the morning the band in the public square began to play, as the signal for the people to assemble. At ten the procession was formed, ready to march to the planting grounds. First: the band under the leadership of Gilbert Gerrish. Second: the children in alternating fours of boys and girls. Third: the adults in the same order; followed by the carriages with the President, the Patroness, Mrs. Bainbridge, Fillmore Flagg and Gertrude Gerrish.

Having reached the grounds, the procession was massed into a square of close columns. The ranks were divided into planting classes of twenty, with an instructor for each class. After the classification, the double quartet of mixed voices, sang a hymn to the forest; the assembly joining in the chorus. As the square broke up, the members of each class, carrying tools and plants, followed the teacher to the particular planting grounds prepared for them. At a given signal, three blasts from the bugle, the work began, and went merrily forward, with much vigor and a vast deal of lively chatter. In just twenty minutes, the planting was finished and the square reformed. The children altogether as a chorus, then gave "An Ode to Growing Trees," which they rendered so sweetly and so effectively, that they earned a great deal of well deserved praise. The order for the return march was sounded—the procession quickly re-formed and returned to the village in the same order in which it came.

A twenty-minute band-concert, given in the large dancing pavillion in the center of the public square, came next, and closed the order of exercises for the forenoon.

An intermission until one o'clock was declared.

Promptly at one o'clock the people were again assembled in the great hall of education and amusement, to hear the oration. The hall itself was handsomely decorated for the occasion, with a profusion of flags and ribbons. The roomy platform was transformed into a garden of verdure, by a brilliant array of ferns, flowers, palms, potted plants and young trees. Seated near the center of the platform were Fern Fenwick, Mrs. Bainbridge, Gertrude Gerrish, Fillmore Flagg and George Gerrish. The latter, as the president of the farm company, in a few well chosen words, introduced General Manager Flagg, as the orator of the day.

Inspired by the cheers which greeted him, happy in the presence of his beloved Fern; yet with all alert, and confident of his complete mastery of the subject; our hero never before seemed quite so handsome as when he began to speak.

CHAPTER XXVIII.

THE ORATION.

"PEOPLE of Solaris, I thank you for the honor of having been chosen as the orator, for this our first Arbor-day Celebration! I assure you, that I am both proud and happy to serve you in that capacity!

"In the beginning, let us consider the art of tree-planting, from the stand-point of an acorn, as being a typical nut or tree-bearing seed, such as I now hold in my hand.

"This tiny nut, with such a smooth hard shell of polished brown, contains a kernel with magical possibilities. Within this kernel, closely packed and safely cradled, lies the embryo oak. So small and so insignificant is this nut, that one may travel for months over land and sea, with the possible ancestor of a half-dozen future oak-forests snugly tucked away in some inside pocket. This, too, without ever once receiving a demand from the lynx-eyed custom officials, for the payment of either import or export duties upon it. Half way round the globe, from the spot occupied by its parent tree, this highly-polished, much-traveled nut, if given the proper conditions, will at once commence the mysterious transformation process, which marks the beginning of the life and growth of another oak tree. This growth, under favorable circumstances, may continue for the historical period of ten centuries. Ministering meanwhile, to the needs of forty passing generations of people. Reproducing itself, perhaps a million times in the aggregate, by the enormous annual crops of acorns it may have borne. What a history of marvels, is the history of such a growth! As it is with the oak, so it is in a large measure, with all other trees which are produced from seeds.

"This fascinatingly mysterious process of passing from seed to plant,—from passive to active life, we have watched with keen interest and growing pleasure, as from week to week, in the seed beds and nursery rows of our tree-garden, it has steadily progressed, under the varying conditions of sunshine and storm. Having reached a suitable size for transplanting, we have this morning commenced the actual work of tree planting, by carefully placing the young trees in the proper soil and location, where they may complete the sturdy growth they have so well begun. The preparatory work, we began some months ago, when as individuals, we selected the three trees, of some one chosen variety, which we especially desired to plant in forest formation, on the occasion of this festival.

"By the months of thoughtful care and attention which we have given to these trees, we have gained a personal interest in them which we cannot lose. In this initiative work, I am convinced that we have wisely established such a broad foundation of general interest in forestry and kindred topics, that sooner or later, it will lead us to a complete mastery of the whole subject. The individual interest thus established, will continue to expand until it embraces the entire tree-family of the world. By constantly adding to our stores of knowledge in this direction, we shall be surprised to find how much we have extended our field of pleasure. In the same ratio, there will come to us a corresponding increase of affection and appreciation for our benefactors, the trees; a solace in the sojourn of life, so generously supplied by Mother Nature.

"The location of Solaris as an experimental tree-planting farm, is particularly fortunate. It possesses a soil and climate which will promote the perfect growth of more than one hundred different varieties of trees. Among these, we find a majority of the valuable timber and nut-bearing trees of the world. Consequently, a very wide field of experimentation awaits our efforts. Let us improve our splendid opportunities so industriously, that a wide spread interest in forestry, may follow and become firmly established in the minds of the people of our Republic.

"By way of an introduction to the general subject, of the importance of trees, as an adjunct to the progress, welfare and civilization of mankind. I wish to relate to you the story of my first great lesson in the seductive lore of forestry.

"Near the beginning of the last decade of the Nineteenth Century, in the year of 1893, it was my good fortune to visit the World's Columbian Exposition at Chicago. I was then a lad of fifteen years, full of boyish enthusiasm, in the enjoyment of my first vacation from the preparatory school, where I was being fitted for my collegiate course.

"I was born and reared on my father's farm, on the broad rolling prairies of Nebraska; up to that time I had never been far from home; as a consequence my knowledge of growing trees was limited to the following fast-growing varieties, which were planted and cultivated by prairie farmers for fuel, fencing and storm-protection. I will name these varieties in the order of their value for fuel and timber. White ash, soft maple, cottonwood and white willow. At a later period I learned that perhaps with the exception of white ash, the timber furnished by these trees, is considered valueless, in the markets of the world.

"Under such circumstances you may imagine my astonishment when I first beheld that wonderfully unique, Forestry Building; with its bristling array of tree-trunk flag poles. Try first to picture in your mind's eye, a building in the form of a parallelogram, large enough to afford two acres of floor-space; with the first story surrounded on every side by a wide, open veranda: with a full length second story one hundred feet wide, rising gracefully from the central roof of the first; altogether, completing a design of exterior so boldly rustic in its general effect, as to suggest the idea of trees and forests at every point; then, you may get the delightfully novel effect, which the architect conveyed to my mind as I approached this curiously fascinating structure. A closer inspection increased the rustic effect of the general design. The main outside walls, were composed of thousands of wide, bark-coated slabs, cut from the choice typical trees of our American forests.

"The wide roof, was in itself an ideal creation; it was thickly covered with curving tiles of rough bark, in alternating layers of the varying kinds, which formed a picturesque combination redolent with the spicy resinous odors of birch, basswood, hemlock and fir.

"Completely encircling the building, with feet firmly planted on its solid stone foundation, rising to the roof through the floor of the veranda at its outer edge, were the thickly planted supporting pillars. These pillars like a long line of watchful sentinels, were placed in trios. The two outside pillars of each trio, were only separated from the middle one by a few inches of space, and were as nearly as possible, ten inches in diameter. The one in the center was much larger and held the post of honor as the flag bearer of its triumvirate. By pushing its way through the roof it became a huge flag pole, fifty feet from base to tip, with a beautiful banner proudly waving from its ball crowned summit. These pillars, both large and small, were bark-coated below the roof. Each one had been carefully selected for its symmetrical straightness, as a representative tree from the different forests of the world. Altogether, they formed a most interesting collection, to which might well be devoted, many hours of admiring inspection, by every lover of trees.

"A wide lattice work of bark-laden tree limbs, of a uniform size completed the charmingly rustic cornice, which, like some endless curtain seemed to hang suspended from the caves of this bark-thatched roof.

"Having sufficiently studied the exterior beauties of this remarkable building, of such arborescent magnificence; let us mount the steps to the broad, breezy veranda. Pausing a moment to inhale the refreshing coolness of the crisp air; and to admire the wave curving sparkle of the blue waters of Lake Michigan, we then pass to the shining portal of richly colored, highly polished woods, which form the main entrance. Here, covering the entire available floor-space, piled high in splendid profusion; we behold the garnered riches from the forests of the world.

"I shall not attempt to describe my varying emotions of wonder and delight, as I wandered for hours through a bewildering maze of the wonderful exhibits, which formed this unrivalled collection of choice woods. As I advanced, my admiration for its variety and extent continued to grow. I began to perceive that, spread out before me, was the opportunity of a life time, which, if properly utilized would prove for me the permanent foundation of an education on the subject of timber, trees and forestry products. With this realization came the resolve, that I would devote time enough to each exhibit, to permit me to examine it in detail, leisurely and carefully.

"The separate exhibits from the States of the Union and from other nations, were skillfully classified and so artistically arranged, as to show in the most effective manner the lovely grain, color and finished beauty, of the different woods.

"All the valuable timbers were represented by three specimens. The first and second, were polished planks displaying the grain-finish, of both radial and transverse sections. The third, a cross section or disc, showing the heart, body-wood, sapwood and bark; the full size of the tree represented. These discs proved by far the most interesting part of the exhibit. To me they were a revelation! They at once introduced me to the individuality of the tree. I could read the history of its life as I scanned the ever-widening circle of annual rings, which, from center to circumference, marked the slow growth of ages, as the tree advanced from infancy to maturity.

"By means of these polished discs, I could touch and become personally acquainted with the precious, the famous, and the historical trees of the world. The mighty teak and deodar from India. The giant mahogany from Central America. The olive of Palestine. The cedars of Lebanon. The ancient oaks of Dodona. The magnificent dye-wood and rosewood of Brazil. The majestic live-oak of Florida. The druidical-oaks of England. The smooth, elastic bamboo, which by its size and strength becomes so useful in house-building, in both China and Japan. The towering spruces and sugar pines of our Pacific Coast. The great elms of New England. The justly famous, white pines of Michigan, Minnesota and Wisconsin. The wonderful spice-woods of Java and Ceylon. The curious soap and rubber trees of Brazil. The tall sugar maples and smooth, symmetrical beeches of New York. The great hemlocks of Pennsylvania. The stately cypress, the royal tulip tree, and the beautiful evergreen white holly, of our southern forests. The highly prized black-walnut of Tennessee and North Carolina. The fruitful, free-growing chestnut, so common all over the United States. Finally, that towering king of all trees, the matchless mammoth redwood of California.

"These redwoods are such veritable giants in size, that the half disc displayed in the California Section, with its thick ring of bark on the rounding side uppermost, stood sixteen feet high. From the huge trunk of this tree came the accompanying plank of such extraordinary dimensions, that a placard proclaimed it the largest plank the world ever saw. This plank was five inches thick, twenty-five feet long and sixteen feet nine inches wide; containing about two thousand feet of lumber, board measure.

"In the Brazilian Section I found a large disc, accompanied by a specimen branch, with the leaves, flowers and fruit of a most remarkable tree. To this tree, the world owes a debt of gratitude for its generous unfailing supply of a rich wholesome food. Almost every child through the sense of sight, touch and taste, is familiar with that peculiar, triangular-shaped, sharp-edged, black-coated nut of commerce, with such a delicious kernel, known as the brazil nut. Very few however, know anything of the tree which bears them, or how they are attached to the branches from which they are suspended. As it is a matter of such general interest to both old and young, I shall take the liberty of devoting a few moments to a brief description of this gigantic tree, which the botanist has named 'The Bertholletia Excelsa.'

"These wonderful trees grow most abundantly in the valleys of the Amazons, and generally throughout tropical America. In size and beauty, they rank as monarchs of their native forests. They attain an average height of one hundred and thirty feet, having smooth cylindrical, beautifully proportioned bodies; which often have the astonishing diameter of fourteen feet, when measured fifty feet above the ground. Like columns in some vast cathedral, these majestic representatives of the vegetable kingdom, raise their massive trunks one hundred feet toward heaven, before they commence to branch out, and to form a medium sized, symmetrical top. At this height grow the flowers and fruits.

"The fruits are globular, with a diameter of five or six inches. Each fruit contains within its black, woody, shell, from eighteen to twenty-five closely packed seeds or brazil-nuts. These fruits, as they ripen, fall from their lofty position. At the proper season they are collected, broken open and marketed by the Indians, who roam through these dark, gloomy, miasmatic forests. The extraordinary abundance of the crop may be measured by the fact, that one port alone on the Amazon River, exports annually more than fifty millions of these excellent nuts.

"Brazil-nuts are largely eaten as a nutritious and palatable food, by a multitude of people in many lands. They yield a generous supply of fine bland oil, which is highly prized for use in cookery, and also for lubricating all kinds of delicate machinery.

"The timber furnished by these fruitful and beautiful trees, is light and durable, easily worked, well adapted to the purpose of boat-building; especially canoes of the largest size. Indeed! I may add as a final tribute to these noble trees, that they are the peculiar product of the American Continent, of which it may well be proud! They have bodies so tall, so straight, so large, so symmetrical, so free from knots, and so easily dug out, that the largest ship used by the hardy and fearless old Vikings of the Eleventh Century, could easily have been fashioned from a single one!

"In connection with the main exhibit in the Forestry Building itself, I visited and examined the magnificent and astonishing timber displays shown in the State buildings of California, Oregon, and Washington. These exhibits were in every way worthy of those three great states of the Pacific Coast; they also served to largely increase the preponderance of the exhibit from the United States as a whole, over that of all other nations combined. The demonstrated extent, variety and wealth of our timber supply, was a matter of profound astonishment to visitors from other lands; while at the same time these things were equally a source of surprise and pride to every citizen of the Republic who saw them.

"After a most delightfully well spent week, devoted almost entirely to forestry productions, I was prepared to sum up my impressions of the significance and value of the knowledge I had gained in my first lesson. It was plain to me that the magnitude and importance of the subject, was but little understood or appreciated, by the average American citizen. I saw that our people were very much in need of some great object lesson like the forestry exhibit of the Columbian Exposition, to make them properly realize the immensity of our debt of gratitude to Mother Nature for her munificent gift of trees to mankind.

"I shall now conclude my story of the Forestry Exposition, by naming from the exhibit the following, as a few of the many things of use and value, which we owe to our benefactors, the trees; things which are so necessary to our comfort and happiness, which in so many ways, affect the progress, welfare and civilization of the world's people.

"Among the more important gifts from the trees I shall place lumber and shingles, used in the construction of houses, barns and all kinds of habitable or industrial buildings; bridges, boats, ships and sailing vessels of all kinds; furniture, fencing and a great variety of farming utensils. Under the head of fuel, I may mention fire-wood and charcoal. In the class of vehicles we have wagons and all kinds of carriages from the stage coach to the pullman palace car. Some kind of lumber or timber enters very largely into the construction of almost every kind of machinery. In the miscellaneous group we find wood-alcohol, dye-wood, medicinal barks, roots and galls; precious gums, resins and all of the spices; the various kinds of excelsior used for packing, bedding and upholstery; wood-pulp and paper, inlaid work, vegetable ivory, and cocoanut shells; the entire series of willow ware, and wooden, or hollow ware. In food products, we are confronted by a most astonishing array of edible sprouts, berries, delicious fruits and nutritious nuts, forming altogether a multitude of things which, in civilized life, we could not possibly do without.

"In considering the impressions conveyed to our minds by growing trees, which inherently possess a sturdy vitality, that can resist the vicissitudes of passing ages; we instinctively recognize them as nature's noblest gift to man. As majestic monarchs, in the empire of plant life, they appeal to us as companions, which become dearer with the associations of each passing year, until love for them becomes a feeling almost akin to worship.

"This worshipful feeling, no doubt, comes to us as a heritage from a remote ancestry. In the days of ancient story, groves of noble trees offered primitive man, nature's grandest and most appropriate cathedrals, for the celebration of his worshipful rites. Is it a matter of wonder, that he unhesitatingly accorded to them, the distinction of being sacred? The emotional nature of this primitive man was a mystery which he could neither understand nor control. Often, he suffered untold tortures from the agonizing perturbations to which it easily became a prey. Hidden in the deep shade of his sacred grove, in his happier moments, the sighing of each passing breeze through his leafy canopy, become to his untrained ear, the whispered blessing of nature's placated God! When the dark pall of the Storm King shrouded all things with a terrifying gloom, the restless moaning of such a mass of writhing boughs, lashed by the fury of the blast, became the angry shriek of the Demons of Destruction, which left him prostrate and trembling in the throes of a paroxysm of worshipful fear. Analyzed, these actions show the result of man's environment.

"By the way of a contrast, and as a testimonial to the planetary growth of man's emotional nature, gained from the ages of progress; let us question modern man as he leans confidingly, in a contemplative mood, against the broad trunk of some giant of the forest. With uncovered head, he muses in silence; he senses a vague feeling of awe for this magnificent specimen of matured life in the vegetable world. With every sense attuned to the overtones and undertones, produced by the vibrations of nature's harp; he catches the rythmic song of the sappy currents, as they swiftly fly to feed the swelling cells, where the building energy of their tiny hearts of protoplasm, ceaselessly changes the elements of soil and sunlight, into the woody fibre of this mighty tree. How beautiful! How like the complicated mechanism of the human body! Wonderingly he questions! Can it be possible, that the pulsing energy of the protoplasmic life of the tree, is identical with that of man, and all other forms of cosmic life? Does each great throb of the planetary heart, re-energize and move in unison, the protoplasmic centers of all forms of life? Who shall say?

"In discussing the peculiar fitness of our present organization, to deal effectually with the question of tree planting, we discover, that in the co-operative association of so many people, we possess a marked advantage over the small farmer, which enables us to treat large tracts of land as a single farm; by devoting all of the rough, stony ground, steep hill sides, unsightly gullies and areas of poor, gravelly soils, to the purposes of timber and fruit culture.

"Harmoniously united, we are financially and intellectually stronger; less influenced or retarded by motives of selfishness and greed; surrounded by conditions of easy comfort; armed with skill by study and experience; and withal inspired by a knowledge of the great necessity for replacing our forests; we are exceptionally well prepared to carry forward this great work, so successfully and to such an extent, that a few decades hence our hill sides and mountains, shall be re-clothed with beautiful forests of much finer trees—all choice timber—vastly more valuable than the original stock.

"By more systematic methods of terracing the steep hills; by close planting of the young trees, with varieties selected by reason of their value for lumber, timber, nuts and fruit; by a judicious thinning out of these young trees so soon as they have grown to a useful size; a profitable crop of timber may be secured each year, with a positive benefit to the remaining trees. This operation may be repeated many times, before a partial replanting becomes necessary. By an extended use of these methods, the excellence of the timber supply may be doubled, while the aggregate yield will be trebled. The landscape will be beautified and permanently changed. Barren, unprofitable hills, and rough unsightly mountain tracts, rejoicing in a new growth of beautiful verdure-clad trees, will become objects of general admiration; while at the same time, the value of these lands, as a source of wealth, will be increased a thousand fold.

"As these forests continue to grow, the shade deepens, the store of retained moisture increases, perceptible changes in the climate are effected; the evils of flood, erosion and drought are checked; the soil made deeper and richer; the rainfall largely increased; the climatic conditions become more genial, and the cooling, drouth-dispelling rains become more frequent.

"The interesting and beautiful process, by which these changes are accomplished, may be briefly stated as follows: With the growth of each year, the area of the leafy surfaces of these forest trees is enormously extended. Measured by the same increasing ratio, many additional thousands of tons of moisture are pumped up and given to the winds in the form of a fine vapor, by the tireless industry of these lovely leaves. This vapor is taken up by the clouds—nature's aerial reservoirs. Soon this treasure of waters thus accumulated, is restored to the thirsty earth by a largely increased rainfall. Autumnal frosts ripen and loosen each crop of leaves; they fall silently to the ground, where they quickly form a thick, soft carpet of ever increasing thickness. Through the action of shade and moisture, the under surface of this carpet becomes a layer of fine leaf mold, which in turn offers rich food for the sustenance of millions of tiny feeding rootlets from the trees of the forest. The closely interwoven fibre of these rootlets, everywhere forms a strong web for the carpet, which firmly holds in place the soft, porous, underlying soil, safely protecting it from the destructive erosion which, especially on the steeper slopes, swiftly follows the dashing violence of heavy rain storms. Gradually this leafy carpet grows in strength and thickness; like some great sponge it sucks up and retains the waters of the snows of winter, with those of the increased rain-fall of summer.

"Thousands of mountain torrents, the beginnings of destructive floods, are thus checked, absorbed and shorn of their disintegrating energies. The garnered waters from this wonderful leafy sponge, slowly percolate through the soil, to reappear in a multitude of living springs of pure sparkling water. From these springs gently flow the tiny rivulets, which in turn become the full streams that gladden the plains and valleys throughout the long scorching months of summer.

"By a close analysis of the beneficial results which follow the annual recurrence of these beautiful processes, we may form a correct estimate of the vast importance of this tree-planting labor, to which this day, we gladly offer our best energies and our best thought. We begin to perceive the magnitude of the blessing which may be conferred on mankind, in general and on the agriculturist in particular, by the continued work of covering our hills and mountains with valuable forests.

"We have discovered from nature the secret of a power that shall enable us to control many of our environmental conditions. We hold the key to the solution of a great problem, which for the past quarter of a century, has puzzled the brightest minds and best thinkers among our statesmen. The problem of how best to control the devastating floods, which each year, with increasing power and violence, continue to destroy hundreds of lives and millions of dollars worth of property, on the farms and in the towns and cities throughout the river valleys of our broad land. For this growing terror, we hold the cure! With the completion of this system of forestry, the floods will disappear. The interests of our coastwise and inland commerce, will be greatly extended and benefitted. Many rivers, with beds choked and obstructed by the unsightly rocks and debris deposited by the annual floods, and for the same reason, dry for many months in each year, will again become navigable. Perennial streams, fed by permanent mountain springs, will serve to keep these rivers with full channels throughout the year.

"The clear water will be free from the lighter silt which now finds its way to the sea ; slowly filling up the river-mouth harbor, and finally destroying the commerce of the city which depends upon it. In this way, every individual, child or adult, who plants a tree, aids directly in the restoring some distant seaport to its former commercial importance; and has proudly earned the right to be placed as an important working member, on the peoples' great 'Committee for Improvement of Rivers and Harbors.'

"Tree-planting, persistent tree-planting, by all classes of agricultural people, offers the only means or hope of checking the wide-spread, calamity-producing floods and erosions, which commenced with the destruction of our mountain forests. The destructive process is accelerated with each passing year. Unchecked, it threatens, a few centuries hence, to rob us of all fertile soil; to reduce our hills and mountains to a dreary waste of bare, sun-scorched rocks: our plains and valleys, to uninhabitable deserts. United action is therefore imperative!

"Other incentives, worthy of our attention, urge us to commence the work. By yielding even one-half of the area of our tillable lands to the needs of forestry, we have all the richest lands left in the remaining half. The productiveness and fertility of these lands is sure to be speedily doubled. The amount of labor required to produce the same crops from the diminished areas, will be reduced one-half.

A most important consideration!

"The third generation of people, after the planting of these forests, will gather from them, such an abundant harvest of nuts, fruits, and valuable timbers, as will more than repay the entire cost of the land and labor required to produce them; leaving a handsome surplus to be devoted to carrying forward the work on a still larger scale; in regions less promising and more remote, even within the borders of the arid lands. With this lesson before us, how can we hesitate or falter in our efforts to successfully carry forward this important work?

"I wish now, to call your attention to the following facts regarding the farms and farmers of our Republic, which altogether offer additional incentives for the speedy adoption of co-operative farming on a scale large enough to admit of timber culture, as the only available source of relief. The significance of these facts has scarcely been considered, by those most deeply interested. The farming lands now owned or controlled by our agricultural people, represent the accumulated capital or savings of a life time; frequently of several generations of the same family.

"A steady decline in the market values of all farm products during the past twenty-five years, has in the same ratio, affected the selling value of the farm to such an extent, that from forty to fifty per cent of its value at the commencement of the decline, has been swept away and lost to the farmer, from the credit side of his available resources. This alarming shrinkage, has in the aggregate, amounted to many millions, yes, billions of dollars! The financial distress which has followed, has correspondingly affected many other industries. It has been the real cause of the forced sale of many fine farms at such ruinously low prices, as to sacrifice at one blow, the savings of a life-time. Each sale of this character serves to depress the market value of all lands in that particular locality. In this way the disaster spreads and gathers additional force.

"A very large number of farmers, who have not as yet been forced to sell their farms, have found themselves so financially cramped, as to be unable to secure the additional lands they had hoped and planned to purchase for their children. What is the result? A most abundant harvest of blasted hopes for the sons and daughters of our American farms!

"Capital in the hands of shrewd people, is always on the alert, waiting for such opportunities for investment. These investors through capital wish to live without effort, upon the proceeds of the labor of others. They seem to understand clearly, that to own land, is to own the services of the people who must have access to the land in order to live. This is why a land monopoly is more to be feared than other kind. For this reason we may well be alarmed, as we note from time to time, the large tracts of land which are being purchased by wealthy individuals, foreign syndicates, home corporations and land monopolists generally, who are quietly operating, while prices are so abnormally low, to obtain such complete control of our valuable agricultural lands, as will enable them in the near future, by a concert of action, to raise prices to such a pitch, that practically they would then be beyond the reach of the ordinary farmer.

"These shrewd, far-seeing monopolists, having obtained control of the lands in question, can dictate such rents to all applicants, as will barely enable them to live. As a matter of fact, it is quite probable that they would much prefer not to rent their lands, because they could save for their own pockets, the wages of a great many workers, for at least five months in each year, by placing five-thousand-acre-farms in charge of a superintendent; who with two assistants, could live on the farm, taking proper care of the stock, tools and machinery, throughout the year. During the seven busy months, beginning about the first of April, transient labor, of the homeless tramp order, could easily be procured to work by the day, week or month, as the needs of the farm might demand.

"The growing competition for even this kind of uncertain employment, would tend constantly to reduce the wages. The danger from this source has boon fully demonstrated during the past twenty-five years, by the adoption of this disposition of their holdings, on the part of a great number of large land owners. The success of the bonanza farm, has proved perniciously infectious. Our small farmers, already in financial distress, cannot hope to compete with such large farms, so recklessly cropped by the monopolist for the largest possible cash returns, without regard for the future condition of the soil. To double the capital invested in five years' time, is the only concern of the investor. Whatever the land will sell for thereafter, is only so much additional profit.

"We cannot close our eyes to these warning facts. They foretell the coming whirlwind of disaster. We may be sure that, if these things are allowed to continue without opposition, long before the close of the twentieth century, our agricultural people will be reduced individually to the abject serfdom of a houseless, homeless day-laborer. At this time it is almost impossible for a majority of the sons and daughters of the farms of our Republic to obtain possession of enough land to enable them to follow in the footsteps of their parents, by devoting their lives to agircultural pursuits. Many of them have already entered the downward path of the unfortunate tenant. Many others have been forced to find employment in other pursuits.

"You ask how can this coming disaster be averted? How can our people be saved from such a hopeless future?

"I answer, by the farmers, united with those who wish to become farmers, coming together everywhere in force; by pooling their issues; by helping themselves; by organizing co-operative farms like this, armed with schools in which skilled workmen may be taught to successfully carry on profitable allied manufacturing industries. Monopolistic farms cannot then successfully compete. With demonstrations, such as we are making here to-day, springing up by hundreds and thousands in each county and state, during the next thirty years, what may we expect? The last remaining serf will have been emancipated. The hopeless tenant and the landless farmer can no longer be found. No one can be induced to toil, for owners of the monopolistic farm. The owners will not and cannot work themselves. The experience of a few unprofitable years will urge them to sell their lands to the co-operators at such prices as they may be inclined to offer. The victory will be ours. A glorious victory truly! But, we must not expect to gain this victory without a severe struggle. In the earlier stages of the movement, the monopolist will soon recognize the co-operative farm as an enemy which must be fought to the bitter end, must be stamped out. To this end they will strive in every way to prevent us from obtaining possession of desirable lands.

"This determined opposition we must expect and be prepared to meet. Forestry will help us to another solution of the problem. As the tree-planting farms continue to multiply, the increased rainfall will cause the area of tillable lands, to gradually extend beyond the borders of the arid lands. Therefore in case of necessity, we may turn to these arid lands for relief. In such an event, the question of forestry becomes an important factor.

"By referring to the tenth annual report of the director of the U. S. Geological Survey, we learn that the arid regions of the United States, comprise the astonishing area of one million, three hundred thousand square miles. This immense region contains more than one-third of all our lands ; a territory much larger than that of the thirteen original states combined. North and south, it stretches for hundreds of miles on either side of the Rocky Mountain Range, that great backbone and water-shed of our Continent. On the west, it covers nearly all of the surface of that vast, broken and irregular basin, lying between the Rocky and Sierra Nevada Mountains. On the east, it occupies that extended and peculiar domain of high plateaus, treeless plains and alkali barrens, known as the Great American Desert.

"From this broad expanse of arid lands, in accordance with the statements of the survey officials, we may choose an area of one hundred and fifty thousand square miles of irrigable lands ; that is lands which may be restored to productive fertility, by means of irrigating ditches along the valleys, and by building great catch basins, near the head waters of a multitude of mountain streams, in which may be conserved, the wasting waters of melting snows and those of the heavy mountain rainfalls combined. At this point we may mention incidentally, that this area of irrigable lands could be largely increased, by covering the available slopes of the Rocky Mountains with dense forests of fine timber. With this accomplished, the annual rainfall would be doubled, while the necessary conditions would be established, which, a few decades hence might yield an annual crop of valuable timber, that would soon repay the entire cost of planting and culture.

"In addition to the last named increase, we may add an area of lands equal in size to the state of Illinois, which are beyond the reach of irrigating streams. We find these lands along the eastern foothills of the Rocky Mountains, and around the borders of the Great American desert. They may easily be restored to fertility, by the skillfully applied labor of a legion of co-operative farms. At varying depths beneath these lands, flow perrenial streams of artesian water. By the spouting, life-giving waters of a vast number of artesian wells, a large proportion of these desert lands can be transformed to an agricultural paradise. The cost of these wells, would be but little more than the expense of the labor required to bore them.

"But, says the objector, are not these mostly alkali lands? Of course they are! And for that reason offer greater possibilities of value! Can they be made to grow wheat, and thus increase the bread supply? Is a question that comes from the mouths of the world's great army of bread eaters, six hundred million strong. Just think of it!

"For reasons which I shall state presently, I hope to be able to show why these alkali lands when properly irrigated, can be made to produce abundant crops of wheat.

"For the past twenty years, leading men of science, who, alive to the importance of increasing the world's supply of wheat; have given close attention to statistics which seemed to indicate that the yield per acre, of the wheat fields in all countries, is steadily decreasing. Decreasing to such an extent as to make it probable, that in the near future, the yield on a large proportion of these lands, will become too meagre to pay the cost of cultivation. A long series of carefully conducted experiments demonstrated the truth of these alarming statistics.

"This discovery lead to a general search for some cheap, available, chemical, compound, which might restore these worn out wheat lands to their former productiveness.

"In an address, delivered at Bristol, England, near the close of the nineteenth century, by Professor William Crookes, president of the British Association for the advancement of science; he says; 'Wheat preeminently demands as a dominant manure, nitrogen fixed in the form of ammonia or nitric acid. Many years of experimentation with nitrate of soda, or Chili salt-petre, have proved it to be the most concentrated form of nitrogenous food demanded by growing wheat. This substance occurs native, over a narrow band of the plain of Tamarugal, in the northern province of Chili, between the Andes and the coast hills. In this rainless district for countless ages, the continuous fixation of atmospheric nitrogen by the soil, its conversion into nitrate by the slow transfiguration of billions of nitrifying organizations, its combination with soda, and the crystallization of the nitrate have been steadily proceeding, until the nitrate fields of Chili have become of vast importance, and promise to be of inestimably greater value in the future. The growing exports of nitrate from Chili at present, amount to about 1,200,000 tons annually.'

"In carefully analyzing this lesson from the lips of Professor Crookes, we discover that the same peculiar climatic conditions which made a Chilian desert so valuable, have been continuously at work in our great American desert for a great many thousands of years.

"For this reason, our uncounted acres of alkali lands, are so rich with stores of this valuable nitrogenous compound, that by proper treatment they may become the most valuable wheat-producing lands in the world. The desert shall become the source of abundance! Under the transforming influence of a generous water supply, forests shall spring up, and fields of waving grain shall flourish around the village homes of a happy, prosperous people! Altogether, we have an empire of these irrigable lands now worthless, awaiting the transforming labor of the homeless and landless, to restore them to productive fertility.

"When thus restored, these lands, at the lowest estimate, will be worth the enormous sum of two billion, eight hundred and eighty million dollars, which in due time may be transferred to the credit side of the wealth account of the nation! Long before this available domain of such vast possibilities has been conquered and reclaimed, the longing desires of all who wish for land, and for agricultural lives, for themselves and their children, will have been most abundantly satisfied.

"In looking over this broad field of possibilities spread so temptingly before us, we are able to discover the importance of the work of tree-planting, which now demands our attention. Strengthened by concerted action, encouraged by new ideas and better methods we become firm in our convictions, that it is an imperative duty for us to continue the good work. We must increase the number of our co-operative farms with their tree-planting schools, until, educated and moved by the force of so many demonstrations, a great majority of the people of this Republic shall demand, that the entire area of the range of the Rocky Mountains within our geographical limits, shall become a permanent, public park; with such a wealth of territory and variety of climate, such beauty of scenic grandeur and magnitude of picturesque proportions, as the world never saw before. This matchless reservation is to be devoted to the needs and uses of forestry, mining, the preservation of its great variety of natural curiosities, and of American Game.

"In addition to this Pride-of-the-World-Park, the people shall also demand, that all of the most available portions of the mountains of the Pacific Coast Range, the Sierra Nevadas, the Alleghenies, the Adirondacks and the White Mountains, shall be reserved by the government, and set apart for the same uses and purposes.

"With the passing of this magnificent domain of mountain territory to the permanent control of the government, would come the beginning of the great public forests; which would clothe with new beauty, cover and protect in the most useful manner, the principal water-sheds of our broad continental possessions. Thus increasing to a degree approaching perfection, the purity and abundance of the crystal flood, that shall flow from a countless multitude of new springs of living water. The volume of water from these springs, shall furnish a supply sufficient to maintain with full channels, a perpetual flow in that net-work of lakes and rivers, that arterial system of fertility and commerce, which variegates and adorns the bright face of our fair land.

"Altogether, in considering the broad scope of this stupendous plan as a whole, we have before us a most important work, which must be accomplished! A work which affects the welfare and happiness of every citizen of our Republic! A work which is in every way worthy of our most earnest and persistent effort!

"This day, we have made a propitious beginning, which augurs well for success. Let us on all occasions encourage tree-planting as a sacred duty which we owe to future generations! A duty which must not be neglected! From this time forward, let us strive in every way to organize a broader, wiser, more powerful movement! Carried forward by the resistless force of an enthusiasm born of a mighty purpose; with strong hands and willing hearts, let us undertake the speedy accomplishment of our chosen task! Let us remember our responsibilities as immortal beings! Let us be mindful that life on this plane of existence is very brief; that an eternity of countless ages lies beyond! Therefore we cannot afford to be selfish! Let us heed the warning of nature's just law of compensation, which declares that in the higher life, selfishness becomes a torment in comparison with which a crown of thorns would seem a coveted blessing!

"In our devotion to this noble work, let us ignore all unworthy thoughts of self interest! Possibly we may not as mortals, live long enough in the material form to reap many of the benefits that are to follow. But, being immortal; and having passed to a higher realm, where we are endowed with a keener, broader, mental, and spiritual vision; lost to the sense of time or physical pain, we may then behold the results of our work, in the increased enjoyment of our children and our children's children; while the centuries, like moments, glide swiftly by and are lost in the endless procession of passing ages!

"Finally, as an additional source of encouragement to continue a work which we may not live to see mature; let us consider carefully the significance of the fact, that he who causes two blades of grass to grow where only one grew before, is counted a public benefactor. Judged by the same standard, he who causes two trees to grow where only one grew before, is a benefactor of mankind, whose good works shall earn for him the blessings of a hundred generations! By the same logic, it surely follows, that the people, who cause a forest of trees to spring from the arid bosom of desert earth, become the distinguished benefactors of the human race, who offer shade, shelter, fuel, fertility and sustenance, to a thousand future generations! They shall be thrice blessed! Having arisen to the demands of a higher life of unselfishness, where the solidarity of all life is recognized as a self-evident truth; they have gathered a sufficient store of love and wisdom to admit them to the domain of causation. Classed as worthy workers in that domain, they are entrusted by nature, with the magical key which unlocks the climatic gate, to her pent up floods of fertility.

"In conclusion, people of Solaris, I leave this presentation of the subject for your earnest consideration until the recurrence of our next annual festival. During the interval, I feel confident that you will all join me in a closer study, of a topic which has already proved one of such absorbing interest,—of such vast importance.

"Thanking you for your close attention, and for the frequent applause, which has demonstrated your approval, I recommend that we do now adjourn, to enjoy the waiting banquet which is to follow as the next order of the day."

* * * * * * * * * *

Great applause greeted Fillmore Flagg at the close of his oration. George Gerrish arose and paid a glowing tribute to the wisdom and eloquence of the orator; after which, grasping him by both hands, he said, "Fillmore, I am proud of you! Solaris is more than proud of the masterful way in which you have treated the entire subject! Your presentation of the theme, seemed to me to be so perfect, so exhaustive and eloquent, that in the future I may not expect to again hear its equal."

The next moment Fern Fenwick came forward, radiant in her loveliness, her beautiful eyes shining with emotions of love and gratified pride. In a voice, whose clear, well modulated tones, thrilled him as no music could, she said, "Nobly done, Mr. Flagg! I knew you would not disappoint me! Your speech was the most lovely poem in prose that I have ever heard! So perfectly charming, that I find it far beyond my best words of praise! In return for such an eloquent tribute, the trees should join in a grateful anthem! You have sounded the key-note; it is the evident destiny of co-operative farming in the twentieth century, to restore these noble trees to their rightful domain."

The banquet, which followed the oration proved a great success. It was really one long, interwoven garland of witty speech and inspiring music, together with the merry jingle and melodious crash of silver and china. The enjoyable zest of the entertainment, was spiced and flavored with the appetizing aroma of an abundance of delicious, well-cooked food. Placed at the head of the first table, our hero and heroine were at all times the center of attraction; the observed of all observers. "A handsome couple, evidently heaven-ordained for each other," was the universal comment.

The dance in the evening, was fittingly chosen as the closing function of this famous festival. In arranging the program, Fern and Fillmore were selected by the floor managers as the leading couple. Inspired by the music of an excellent band under the leadership of Gilbert Gerrish, the assembled guests with the vigor and enthusiasm of youth caught the prevailing spirit of merriment, and gave themselves up to the fascinating movement of musical measures. Lost in the charm of the mazy dance, the merry-makers noted not the flight of time. The last number on the program came all too soon for them.

Dismissed by George Gerrish, the people of Solaris left the hall in a joyful mood. They declared with one accord, that the day of the tree-planting festival, had proved the happiest one on the farm.

CHAPTER XXIX.

THE STORY OF GILBERT GERRISH; OR, THE STRENGTH OF THE WEAKEST UNIT.

To Gilbert Gerrish the day of the festival was one long to be remembered: a day so laden with enjoyment for him, that all consciousness of his affliction was blotted out. His musical genius was free and unfettered. In such a mood, the music he drew from his violin was more wonderful and entertaining than ever before. Fern Fenwick was astonished and delighted. She soon became so much interested, that at intervals between the dancing, she came upon the platform to engage him in conversation. Grateful for such marked attention from the distinguished patroness of the farm, the natural shyness and reticence of the young musician, was quickly dispelled. To Fern, it was remarkable how eloquently and interestingly he could talk upon almost every topic she chose to introduce. On the subject of ethical, social, inventive and educational work, as exemplified by the different phases of club life at the farm; Gilbert was at his best. He spoke with such enthusiasm and perfect knowledge of details that Fern Fenwick was profoundly impressed. She then and there determined, at the first convenient opportunity, to have Fillmore Flagg relate to her more in detail, the many incidents connected with his farm life, and how this interesting boy had managed in so short a time, to make himself such a universal favorite with the farm people, both old and young.

That night before retiring, Gilbert told his mother

in confidence, that Miss Fenwick was the brightest, most beautiful and most lovable woman he had ever met. "Tell me truly, Mamma! Do you think she is really in love with Mr. Flagg? I hope it may be true! For I know he deserves to win the love of the best and most charming woman that ever was born!"

While this confidential interview between mother and son was in progress, Fern and Fillmore were speaking of Gilbert in such a way, that if overheard by Gertrude Gerrish it would have stirred the pride in her mother heart.

"I declare, Fillmore!" said Fern, "to my mind that clever lad, Gilbert Gerrish, is one of the most astonishing products of Solaris Farm! You have promised to tell me the story of his life here on the farm. I am now ready to hear it. At the festival dance I had an opportunity to engage him in conversation, and the good fortune to so win his confidence, that he could talk to me without embarassment. It was then that I discovered what a brilliant intellectual prodigy, eloquent talker, skilled musician, and cultured artist he really was. There is something mysterious about his strong, intellectual, spiritual nature, which has aroused my interest in him, and my sympathy for him, to a degree that is very unusual for me. The more I know of him the more I wish to win his friendship.

"What a terrible misfortune, that he is so afflicted by the deformity of that spinal trouble! I cannot help picturing him as possessed of a physique in harmony with his glorious intellectual and spiritual unfoldment. How naturally then, he could win the love of some equally gifted, noble woman. How happy they could make each other through the passing changes of a long and useful life. Aside from my speculative fancies, I do wonder what the future has in store for him? How bravely he bears himself! He does not seem inclined to be gloomy or misanthropical under the burden of his misfortune!"

"I think, my dear Fern, that my story will unravel the mystery. I am delighted to find that you have already become interested in Gilbert, and have discovered so many of his good qualities! I can assure you that he is worthy of your sympathy and friendship! He is a noble fellow! Richly endowed, with a remarkable, intuitive, spiritual nature! His enthusiasm, persevering efforts and ingenious devices, have contributed much towards the success of this co-operative farm. The value and variety of his especial work in the department of experimental farming, has proved his extraordinary ability, and justly earned for him the title of the 'wonder worker of the farm!'

"On account of Gilbert's frail form and sensitive nature, it was deemed wise by his ever watchful parents, to give him the protection of an isolated home life. For this purpose, a cozy cottage was built in the center of its own grounds, some distance away from all other buildings. This cottage was charmingly fitted and furnished in such style and taste as would satisfy the artistic ideas of this domestic trio, and at the same time, afford quiet, retired, spacious rooms, for Gilbert's musical and other studies. Rooms where violin and piano practice, at any hour that might suit his fancy, could disturb no one.

"Referring to that haunting desire which impresses you to picture Gilbert as possessing a magnificent physique, in harmony with his brilliant, mental and spiritual unfoldment; I accept it as another proof of the growth of his spiritual body to the beautiful proportions you seem to see. All psychics who come within the radius of his powerful, spiritual aura, sense or see this strong symmetrical body. His affectionate and emotional nature is beautifully developed. No one can appreciate the graces and charms of a refined, beautiful woman more keenly than Gilbert Gerrish! Yet, I know, that in this life, he does not for one moment, even dream of a possible marriage with any woman. He is loyally devoted to his spiritual ideal!

"For many months, I have been to Gilbert a trusted friend and confidential companion. In this capacity, I have learned his story of the hidden romance of his young life. This story I will repeat to you as an illustration of the high order of his boyish character. It cannot fail to increase both your admiration and your respect, for this youthful devotee at the shrine of love.

"When Gilbert was ten years old, while attending school at St. Louis, he became acquainted with Rita Estelle Ringwood. She was in many ways a remarkable girl; only two months younger than Gilbert. Tall and straight, with a well rounded figure, already as large as a maid of fourteen, Rita gave promise of an early development into a lovely woman. With a large, finely formed head, crowned by a luxuriant growth of soft, thick, wavy, chestnut hair; a smooth, creamy complexion, pleasing features, firm mouth and well rounded chin; large, full, soft, brown eyes, unusually expressive; a strong, well turned white throat and neck, symmetrical shoulders, perfectly formed hands and feet; and a well poised, graceful carriage, she appeared to Gilbert as some divine creature. From the first moment of meeting, a strong bond of mutual attraction drew them together. If kept long apart, both became nervous and restless. When again united, they were quickly at peace with themselves and all the world. By a strange coincidence, as it transpired ; Rita's parents lived in a house just across the street, almost in the front of the one occupied by the Gerrish family. Through the children, the parents soon became intimate friends. As Gilbert had never cared to play with boys of his own age, either on the streets or at school, it was natural under the circumstances, that he should devote himself entirely to Rita, as the only congenial playmate he had ever known. Very soon, as a consequence, the twain were almost always together, either in one home or the other. They read or studied from the same book, often pausing to discuss some question of more than usual interest. In music, they had the same tastes, the same predominating passion for it. Gilbert soon taught Rita to use the violin; while Rita in turn taught Gilbert to play the piano. Each could then alternate, in playing violin accompaniments to piano music. Much practice soon enabled these artistic children, to render such duets with thrilling effect. In so delightful an occupation, hours passed swiftly by. A series of selections were chosen for evening concerts. The parents were called in to enjoy them. In the eyes of the parents, both children were manifestly helpful to each other. Rita never seemed to notice Gilbert's misshapen body. She evidently responded, only to impressions emanating from his more perfect and dominant, spiritual body. Gilbert was conscious of this fact, and always seemed at ease in her presence. As the months flew swiftly by; these strange children grew more devotedly fond of each other. Three summers had witnessed the growing together of these two harmoniously attuned souls.

"The day following Gilbert's thirteenth birthday, he was depressed by some overshadowing cloud of sadness. He could not explain it, nor, could he throw it off. The sequel came the following week, when a great wave of pestilence, in the form of malignant typhoid fever, swept over the city. It claimed Rita as one of its first victims.

"Heart broken! Rita's parents hastily returned to New York, where, surrounded by early associations, they vainly and hopelessly struggled to forget their terrible bereavement.

"To Gilbert, the shock was frightful! His parents, George and Gertrude Gerrish were alarmed. They feared for his life! He wandered about with dry, staring eyes, like one in a trance. He could not weep! For days, he could neither eat nor drink! At last, came the crisis! Reason seemed about to leave her throne! Then it happened, that Gilbert grew strangely calm and hopeful.

"In a few short days the improvement was magical. His beautiful eyes shone with the fires of new inspiration! Questioned by his parents, he assured them that Rita still lived. He knew that she was not dead! Clairvoyantly, he had seen her, more beautiful than ever. Clairaudiently, he had heard, over and over again, the sweet familiar tones of her voice. All this through his own mediumship and more besides. Controlling his hand and arm, in her own identical hand-writing, she had written to him long messages filled with loving consolation, bidding him look hopefully forward to a happy reunion in the land of the spirit, the home of the soul! Almost nightly in dreams, she came to him, when for happy hours they were again united in the enjoyment of the old familiar companionship, so dear to his waking memories.

"Through Gilbert's mediumship, his parents became spiritualists. This happened some months before I visited them in St. Louis, on my first trip west, from Newburgh. Some months later, the family came to Solaris.

"In a recent conversation, speaking to me of his life work, his hopes and his ambitions, Gilbert said: 'Fillmore, I know that my life here will be short. I know that I have a work to do here on this farm, for the future benefit of my brothers and sisters in earth life. I know that in spirit life, Rita waits for me to join her, when that work is finished. I now realize that swiftly passing days, weeks, months and years, are precious portions of time which I must improve to the utmost. I know that this primary school of life has many useful lessons, which I must master as quickly as possible. I know that the sooner they are mastered, the sooner I shall be prepared to enter a higher class in spirit life. I know that as a spirit, in that land of golden sunlight, freed from the burden of this unsightly prison of flesh, I shall be clothed in a spiritual body as symmetrically perfect as my highest ideal can picture. I know that thus clothed, and crowned with the perpetual youth of the spirit; I shall again be united with my darling Rita, never more to part. Together, in obedience to the law of an infinite love, we shall go hand in hand, up the paths of wisdom which lead to the summits of the hills of everlasting progress. I know that during my sojourn here, when I am weary and most need the healing balm of her presence, my Rita can come to cheer and help me. Knowing all this, life is full of promise! I have no time to be sad or lonely! The world is bright! I am ambitious to make its people my friends, by creating for them, better and brighter conditions for the enjoyment of life.'

"This, my dear Fern! is the romance, which like some secret charm, Gilbert wears in his heart. His armor against all evil! The bright star of his ambition! The beacon light of his hope!"

"The romance is indeed a most extraordinary one! The story is exquisitely beautiful! Its pathos fills my heart with both joy and sadness! In the development of his mediumship, following his bereavement, how like my own, has been his experience! This explains my sympathetic desire for his friendship. What a noble fellow he is! I shall be proud to claim him as my friend! Now Fillmore, you must tell me of his work for the farm. I am anxious to know more of the peculiar methods of this inspired genius."

"Very well! In the center of the large garden at the rear of the Gerrish cottage, is a roomy workshop, built for Gilbert's sole use and occupancy. Alone in this shop, he has mapped out for himself such a course of study, experimental work, and industrial amusement, as might suit the fancy of his swiftly changing moods; or conform to the passing whims of his busy brain. To the combined interests of Solaris farm, he is intensely devoted. To keep a realistic picture of the farm always in his mind, he has drawn an immense map, large enough to completely cover the wall space on one side of the shop. He subdivided, colored and named the subdivisions on the map, after a bold, brilliant scheme of his own. The result is a matter of astonishment to all beholders. The map seems to possess some charm of attraction, which no one can explain. On each subdivision from time to time, Gilbert has tacked cards filled with finely written notes, setting forth from his own standpoint, a history of the subdivision, its peculiarities, and capabilities of the different soils; character of crops and fertilizers, together with such suggestions for perfection or improvement, as his thorough knowledge of chemistry might determine; or his keen, analytical, observation of the crops produced, might indicate.

"This map of itself, is a most valuable work; involving an immense amount of intelligent, skillful labor; also much study of chemistry, and of horticultural and agricultural authorities. As an indication of our appreciation of its value, this map has been taken as a suggestive model for the completion of those made and kept by the clerical force employed in the farm office.

"On the south side of his shop, two large doors open into a roomy, glass-roofed hot house, containing a very unique collection of potted plants, which, under the skillful hands of this young enthusiast, are undergoing the different stages of experimental treatment, such as he may deem necessary, to prove or disprove his many pet theories or fancies, in regard to care, growth, insect enemies, and to application of electric light, sun light, heat, moisture and fertilizers. Each plant bears a fruitful crop of cards, giving a summary of results and conclusions. Each one of these cards may contain, in skeleton form, the subject matter of a brief essay, brimful of valuable suggestions and interesting statements. Sooner or later, these essays, signed 'Experimenter,' are liable to find their way into the contribution box at the door of the Press Club.

"Gilbert's collection of birds and insects, forms another interesting feature of his industrial museum. These collections were made, arranged and classified, in order to afford opportunities for making a careful study of the insect enemies of his plants, and also to discover what birds were most destructive to the different insects. The birds he kept in cages; the insects in glass-covered boxes.

"The care of these things, and the time and labor necessary to collect, classify and arrange them, would to most people, prove a grievous burden. To Gilbert, it was simply another mode of recreation and amusement. On the live insects, he tried the effects of such chemicals as might destroy them without injury to the growing plants. To his caged birds, Gilbert fed his bugs, worms and moths, carefully noting the kinds they most eagerly swallowed. His conclusions were always briefly written out. They proved a perfect mine of valuable information, to be used in perfecting better methods for farm culture.

"Aside from this kind of work; in the departments of his shop devoted to experiments with clays, mica, soils, minerals and the various powers, attractions and affinities of electricity, his constructive ideation and inspired mentality, always gave him an excellent crop of good results. Altogether, such superior work, carried forward in his own unique way, has added many hundreds of dollars to the annual income of the farm. In the department of experimental farming, as I have before stated, his work has proved most brilliant and helpful; generally leading to the adoption of many improved methods for successfully selecting, planting and growing these new crops.

"Considered as a whole, such a variety of valuable contributions have convinced our people, that physically speaking, one of the farm's weakest units, under the fostering development of co-operative organization, is capable of becoming one of its most valued productive workers. The wonder of it all, is, that Gilbert is able to accomplish such important results, while following a scheme he has devised as a source of personal diversion!

"Turning to Gilbert's intellectual, artistic and esthetic life, we discover that this gifted boy finds the same source of comfort and amusement in his devotion to the art of music. In this branch of accomplishments, you, my dear Fern! have had occasion to observe how important a factor he has become, in organized social life at Solaris. He is such a general favorite, that without an effort, he has been able to so impress the strong individuality of his noble character upon the minds of our farm people, that the effect for good has been truly wonderful!"

"This is exceedingly interesting, Fillmore! How charmed I am with your completed story of this marvelously gifted boy! All that you have told me about Gilbert, only seems to confirm my previous convictions, that he is really one of the most astonishing products of Solaris farm! No wonder he is such a general favorite! He has nobly earned the title! With such intelligence and genius, possessed, embodied and expressed by its weaker units; is it any cause for wonder, that the success of Solaris as a co-operative colony, is so pronounced?"

CHAPTER XXX.

OUR HERO AND HEROINE DISCUSS AGRICULTURAL STATISTICS.

ON THE day following the festival, we find Fillmore Flagg in the office of the farm, going over the books of the company with Fern Fenwick. To most women, such a task would soon prove unbearably monotonous and tiresome. However, she neither grew restless or inattentive. At all times on the alert to note each new point of interest; her questions on every subject indicated a remarkably intelligent conception of the general plan of the work. Finally, having satisfied herself that she understood the status of the farm well enough to enable her to propound her list of queries in the proper order, and in such a manner, as would most successfully bring to her the information she wished to obtain: with note-book in hand, she commenced by saying: "Now Fillmore, I am ready to take up my series of questions about Solaris, which you have kindly consented to answer. I promise in advance to be good; to try to refrain from untimely interruptions, by asking a host of irrelevant questions at inopportune moments!

"First, I wish you would tell me just what is represented by the one thousand shares of capital stock, of the Solaris Farm Company?"

"The corporation, as you know, is so limited," said Fillmore, "that the land cannot be sold, and the stock can only be sold to the Company; nevertheless, the original cost of the land is covered by the stock. The entire capitalization of $250,000, which I think will fairly represent the financial status of the farm at the end of the first five years, is divided as follows:

Purchase price of land	$ 32,000
Improvements	68,000
Buildings	100,000
Live stock, equipment and machinery	50,000
	$250,000

Of the last named item, about $25,000 is estimated for machinery. However, this amount does not fully represent its real value. In many instances, it only gives the actual cost of the raw material used in construction. This capitalization does not seem so large, when we consider the small individual holdings. Having a par value of $250 a share, we have only $500, in the two shares, for each one of the five hundred co-operators. I think it has been wisely determined by a majority vote, that as the resources of the farm continue to develop and mature, the increase of profits shall come to the individual stockholder in the shape of larger wages, instead of by dividends on stock. Although this is not a money-making institution, and was not so intended from the beginning; a fact properly emphasized by the foregoing. Yet, by the way of arriving at some estimate of its future value, I feel safe in predicting, that, if the stock should be offered in the markets of the world, and dividends declared in the usual way, twenty years hence, these certificates of stock would be worth $1,500 per share. In other words, would have doubled in value six times during that period."

"Judging by what I already know of the farm and its resources," said Fern, "I quite agree with you in this view of the matter.

"In considering the future needs of such a large

number of co-operators, which in ten years may be increased by pensioners and children, to one thousand people; do you think this farm is large enough to meet the demand?"

"For the purpose in view it is ample," said Fillmore. "Operated in connection with so many allied industries, I think a farm of 5,000 acres would be sufficient. That would be ten acres for each one. Here in Solaris, we have 12 8-10 acres of land for every adult member of the company. By carrying the process of intensive farming to a very high state of perfection; Prof. Grandeau, at Capelle, France, has actually demonstrated, that it is possible to grow 8 1-2 bushels of wheat—one man's bread food for the year—on one-twentieth part of an acre of land. Armed with so many advantages, with better conditions, superior methods, and more intelligent workers; I feel sure we can easily accomplish here, all that Grandeau has done in France, and more. Besides, you must remember, that we shall have the additional support of quite a large number of profitable industries, to help us in meeting the demands of an increased number of consumers."

"That sounds logical and reasonable," said Fern. "I now remember, that while traveling in Europe with my father, gathering agricultural statistics: the Capelle experiments were brought to our attention at that time, as worthy of careful consideration. I am greatly pleased to know that you are already familiar with them. To continue the subject, I wish to say that I am much impressed with the outlook for intensive farming at Solaris. Aided by the wonderful power of applied co-operative thinking, combined with your careful and comprehensive system of book-keeping, which embraces every field and department of the farm! I believe that ten years hence, you will be able to give to the world, some very valuable statistics on the whole subject of farming, both intensive and diversified.

"I have noticed with an unusual degree of interest, the apparently lavish use of electric power in operating the factory works and farm machinery. I am really quite curious to know just how it is generated."

"That is a very large question!" said Fillmore. "At different times since the commencement of our work, we have used three methods for generating electricity. First, the old fashioned steam dynamo. Second, the direct conversion of coal into electricity. Third, the gathering of great quantities of this subtle force from the atmosphere, through a certain vibratory action, set up by intense concentration of the sun's rays. As a result of a vast deal of co-operative thinking and careful experimentation; the last named process, has been so perfected and cheapened, as to entirely supersede the first two. The powerful batteries of Solaris concentrators, which you see around the power-house, and at various points on the farm, are important factors in this work. I confess, that I am rather proud of the remarkable success, which we have achieved in this line of invention. When I gave a title to the farm, I had a premonition, that solar heat and force would be so successfully harnessed to both industrial and agricultural work, that the suggestive name of Solaris, would soon become as famous, as it was fitting and well earned.

"In applying this power to all kinds of farm and factory work, we have succeeded far beyond my most sanguine expectations. With a plant almost entirely built by our own co-operative labor, we are able to generate an abundance of cheap power, which can be easily and safely conducted to the most distant portions of the farm. This power is readily available at any desired point, and for all kinds of work ; becoming the magic motor by which we operate trains of trolley cars, for handling grain, hay, corn and all heavy crops ; great gang-plows, rollers, harrows, cultivators, planters, drills, reapers, threshers and motor wagons ; all so perfectly constructed and so easily controlled ; that with them a woman, fittingly dressed and gloved, protected from the heat of the sun by a canopy, comfortably seated on cushions and springs, may accomplish the roughest and heaviest kind of farm work, without fatigue or discomfort. In fact, our women soon find it the most delightfully, fascinating work on the farm.

"In connection with such a powerful motor, a single person, operating one of these improved agricultural machines, can do an amount of work in six hours, which under the old system would require ten hours of severe toil by six men and twelve horses. Of course, such machinery can only be produced and operated by large co-operative farms like this ; with a carefully chosen force of co-operators, who are thinkers as well as workers; who are intellectually, physically and socially prepared to invent and construct machines that are perfectly fitted to do this particular kind of work."

"Really!" said Fern, "this is as interesting as it is remarkable ! This sun-generated force, this magic motor, so perfectly adjusted to agricultural work, under the test of practical use ; which has proved so easily controlled ; together with the tireless host of wonder-working machines, which this force has called into being; is truly a marvel worthy of the twentieth century!

"Tell me, Fillmore! Why is it that these things have not been done before?"

"There are many reasons. I think I can give you the principal one. From a remote period of time, a large majority of the people of this planet have gained a living by following agricultural pursuits. Bowed down under the weight of severe toil, hopeless under the pressure of a belief, that labor was a curse which they might not seek to escape; confined by ignorance to a narrow sphere of action, which kept them from looking upward and outward; it is not strange, that so many passing generations of these people, should never once dream of adopting a series of progressive changes for the betterment of their condition.

"Such people were incapable of understanding, that, in order to secure the best and most successful results from agricultural work, it requires a systematic application of the highest order of brain work: that this brain work, must inspire a harmonious collection of trained, muscular workers, operating under the most favorable conditions. By the way of a contrast, how helpless were the lives of these farmers! As a rule they worked under the most discouraging conditions, distrustful and envious, uneducated and narrow minded; how could they be prepared to comprehend that basic law of progress, which is embodied in the idea of unselfish co-operation?

"For these reasons, co-operative thinking and co-operative farming, have not heretofore been successfully combined. Here and now, in the first decade of the twentieth century, a few unselfish souls, the advance guard of the coming army, responding to the pressure of progressive evolution, have risen to such intellectual heights as has enabled them to discover, that by the aid of a harmonious union of thought and labor, a collection of people, working the soil unselfishly together, can easily attain results which, the most brilliant individual effort, armed with the wealth of a millionaire, could never hope to accomplish. Inspired with this idea, the people of Solaris, as pioneers in the work, are striving earnestly to demonstrate the absolute success of co-operative farming."

"What I have seen with my own eyes, I know as a verity!" said Fern, enthusiastically. "Therefore I feel like shouting in the ears of our people: Well done, good and faithful servants in the cause of progress! The victory is already won! It is yours!

"Your explanation of the cause of the late coming of practical co-operation in agriculture, appeals to my mind, as a very clear one. That the ignorance and selfishness of the individual, has from the beginning, proved the real obstacle, is now quite plain to me.

"However, returning to my list of questions. How is it, that the fields and cultivated grounds at Solaris, are so free from weeds?"

"Ah!" said Fillmore. "The answer to that question, is another argument in favor of co-operative farming. Weeds have always been counted by farmers, as among the worst of the pests which they have been obliged to contend with. Under the most adverse conditions, weeds will grow, flourish, and ripen an appalling quantity of seed; where all useful plants will languish and finally perish. To keep them down, is a task which requires a great deal of hard work. To destroy them, root and branch, is a problem which has occupied the minds of our people for the past thirty months. After much thoughtful work, we have reached a solution.

"During the period of frost, from the first of December to the first of March, the weedy ground is thoroughly stirred several times. After each stirring, the ground is swept by a broad stream of concentrated heat-rays—both light and dark. These rays are generated by a number of batteries of Solaris mirrors, or great sun glasses. This operation soon warms the ground and causes the weeds to put forth a tender growth. After such a growth, a week of frosty weather kills it down. This process is repeated until the weeds are all gone. When the necessary frosts do not appear, or when the work is carried on during warmer weather, a scorching from the sun glasses, kills the weeds even more effectively than frost. In this way the cultivated ground on the farm, has been entirely freed from weeds. As a result, the yield of crops has been largely increased, while the labor of cultivation has been correspondingly reduced. That back-aching work of hoeing, has been almost entirely dispensed with. Machine culture does the work.

"The great advantage gained by cropping soil free from weeds, is most apparent in case of wheat culture. In such soils, the wheat can be deeply sown by the drill, beyond the reach of predatory birds. This develops a strong root-growth in the young plant, which as a consequence requires more space. To meet this demand, care is taken to have the drill-rows made one foot apart—running north and south. These wide rows allow free access of air and sunlight to the soil, which may then be cultivated. Under the old system this space would be full of weeds; therefore impracticable. This gives the young wheat a chance to spread out, to send up from twenty to forty stout stems from the root-system of a single grain of seed. The growing stems become more sturdy, bear larger heads, heads with more and larger kernels, of heavier, brighter wheat. With this culture, the yield is increased one-third—many times one-half—and the quality wonderfully improved. Fully one-half of the usual quantity of seed is saved.

"By repeating this method for a few years, carefully choosing the seed for each planting from the best kernels borne by the largest heads, the ordinary wheat-crop, without extra fertilization, may easily be doubled two and one-half times; while the quality of the entire crop is raised to the grade of extra fine, which will readily sell at fancy prices for seed wheat. The net gain, is a large cash balance in favor of cultivating a weedless soil. What is true of wheat culture in such soils, is true in a large measure with most other crops; more especially with corn, cotton and all kinds of garden crops."

"Stop a moment, Fillmore!

"Did I understand you to say that these immense discs, these mammoth, weed-scorching mirrors, were made here at Solaris? How can such expensive things be made, for a price that would allow so many to be used?"

"Yes, these concentrating mirrors and burning glasses combined, are the product of the inventive genius and skillful work of our people. A combination of brain and muscular work so successful, that these discs, although they are of such great size and weight, are quickly and cheaply made from thick plates of flat glass, which we manufacture from our abundant supply of excellent sand! The quality of the glass in these plates is of the best; clear, soft, and tough, just the kind that will most readily take the proper concave and convex surfaces, when treated by the evenly applied heat of swiftly revolving electric brushes. With plenty of strong machinery to handle these heavy plates, a few skilled workers, can with ease, soon transform them into perfect, lense-shaped discs. Similar discs, made by the slow, tedious process of nineteenth century methods, would cost many thousands of dollars for each one."

"You have answered my question both briefly and perfectly! I recognize in these great mirrors, a swift, wonder-working agency, that shall make possible a new system of farming; which means, in the improved conditions for mankind that must follow, a revolution in social methods, calculated to bring them quickly into harmony with a rate of progress demanded by the twentieth century.

"I will take up another question. It is in connection with the large amount of cultivated ground devoted to vegetables. How do you manage to make it profitable to grow such a quantity of perishable things?"

"That is another important question, which will require an answer so lengthy, that perhaps you may grow weary before I have finished. However, I will try to be brief. During the past year, we have taken from the ground devoted to vegetable growing, more than 100,000 bushels of cabbage, cauliflower, onions, beets, mangel-wurzel, carrots, parsnips, salsify, potatoes, sweet-potatoes, cassava, turnips, kohlrabi and artichokes. The best part of the story is, that this heavy crop has proved profitable, to a degree far beyond our expectations! As a rule, this class of vegetables, so heavy and so perishable, cannot be profitably grown in large quantities, except in locations near a large market town. This advantage, Solaris does not possess. To overcome this difficulty, was an additional task, which must be conquered, by the allied forces of co-operative thinking and co-operative working. In the solution of this puzzling question which was finally reached, the great mirrors and burning glasses of the Solaris concentrators, were again called upon to play an important part.

"The first necessity, was to reduce the weight of the vegetables, and at the same time, to arrest all tendency to decay. The second was to protect them from the attack of insects, by placing them in neat, strong, insect-proof packages.

"A large curing establishment was built and equipped with machinery; most of which was made at Solaris, from especially devised patterns. Convenient trolley lines, connected the curing-house with the fields. The vegetables, crisp and fresh from the ground, were quickly brought to the washing machines, on trains of cars laden with shallow trays, which permitted them to be swiftly handled without bruising. In these machines, they were thoroughly cleansed, scraped, and freed from tops, rootlets and imperfections. This process complete, they were placed in trays on traveling carriers, which delivered them to the dicing machines. In the dicing machines, they were soon reduced to inch-cubes.

"In passing from these machines, the cubes fell

on traveling screens of fine wire, which formed the first of a long series of drying rollers. The drying rollers, on the way to the packing rooms in the large store-house, passed through a long system of sheet-iron conduits, which were well heated by the concentrated rays of the sun from the mirrors and sun-glasses. So well did the drying rollers do their work, that by the time the cubes had reached the store-house, and were delivered by the elevators into the storing-bins in the packing house, they were reduced to a dry, hard kernel. They had lost three-fourths in bulk, and about the same proportion in weight.

"The funnel-shaped bottoms to the storing-bins were so arranged as to be above the long rows of packing tables. A series of graduated spouts, delivered the cured vegetables to the packers, who, standing or sitting as they might prefer, could, with but little effort and much speed, fill the prepared boxes with the little cubes.

"These boxes, of a uniform size and shape, were made from thick layers of heavy straw-paper, made stiff and firm under high pressure. The farm in manufacturing them, was able to utilize large quantities of surplus straw from the grain fields, which could not be used as forage. In the corners of the boxes, between layers of paper, while they were being molded into shape, were inserted small, triangular pieces of wood. These bevel-shaped strips were cut six inches in length, just the depth of the boxes, in which they served as upright cornerposts. The shallow covers fitted each box with a telescope joint.

"In the process of box-making, the layers of paper were saturated with a chemical, germicide solution, which made the boxes insect-proof; yet, which would not odorize, nor in any way injure the contents. In the process of packing, each box and cover was lined with thin sheets of parafine paper, as an additional guard against moisture. When the boxes were filled and sealed, they were strongly coopered, by adding four thin laths of strong wood. These laths, one-eighth of an inch thick, two inches wide, and just the length of the box; two at the bottom, and two at the top, were securely nailed to the cornerposts; thus completing a package which was cheap, strong, light, durable, rodent and insect-proof. With a capacity of a half-bushel, it weighed only five pounds. Filled with cubes, the gross weight was but thirty-five pounds. An ideal package, which could be piled high in transportation or store-house without injury; the upright corner-posts taking all the pressure.

"The half-bushel or thirty pounds of dried cubes in each box, represent two bushels of fresh vegetables. Cured and packed in this way, they reach distant markets, sound, sweet, clean and nutritious. No waste, no worms, no musty smell, no decay! Frost cannot hurt them, heat preserves them! For long voyages, army and navy use, mining, lumbering, and hunting outfits, they are simply invaluable! For all classes of consumers, they are cheaper, cleaner and more wholesome than the ordinary stale and wilted vegetables, for sale in the city markets! We have named these cubes, 'Solaris Vegetable Concentrates,' a title which we have copyrighted. The packages readily wholesale at 75 cents, to be retailed at one dollar. At these prices, they yield a handsome profit to the farm.

"Last year we placed hundreds of sample packages on the general market, which soon proved the excellence of the goods, and later brought heavy orders for this year; even more than we can fill, for many of the varieties. A valuable hint to us, that we must devote more ground to growing those particular kinds.

"Our 'Solaris Mixture Concentrates' are almost equally popular. We also have a growing demand for our 'Solaris Stock Food,' which we put in cheaper packages, to wholesale and retail at 50 and 75 cents. This mixture is made up of equal proportions of dried cubes of potatoes, carrots, cassava, and mangel-wurzel. It has proved the acme of a healthful, fattening stock-food; especially beneficial in counteracting the evil effects of heavy grain-feeding; or in cases of emergency, to take the place of forage or cut-straw food.

"In a weedless soil, much of the heavy labor of growing vegetables is eliminated. In curing and preparing them for market in this way, a great amount of light, pleasant work, is available for our women co-operators. Considered as a whole, this vegetable scheme is one of the notable achievements of Solaris farm, of which the members of the company are justly proud."

"This is surely a most excellent work! It is a clear demonstration of what important results may be attained, by the application of thinking to agricultural work. In this instance, the lesson of your brilliant success, impresses my mind as a most convincing argument in favor of co-operative farming. I feel sure that it will appeal to the multitude with the same force. It is but another illustration of the old saying, 'Nothing succeeds like success!' A few such examples will serve to overthrow the prejudices of a thousand years! They will win for you a host of followers in the cause of co-operative farming.

"Now Fillmore, let us consider another matter. At the time we made our tour of inspection, my attention was attracted to groups of oddly constructed barns, scattered here and there about the farm. What are these buildings, and for what purpose are they used?"

"Those are curing-barns. They mark another wide departure from the usual methods of ordinary farming. For many years it has been a ruinously, wasteful custom with farmers, to allow their crops of corn, grain and hay, to stand in the fields while curing. All, subject meanwhile to the destructive effects of storms, dews and all kinds of adverse weather, which as a rule, destroyed much of the crop, and reduced the remainder to the condition of an inferior grade.

"By the use of these barns, we are able to inaugurate an entirely different system, which succeeds admirably. These barns, located near the grain fields, are constructed with strong frames. They are both tall and wide, and so anchored to their foundations as not to be overthrown by high winds. Each roof is supplied with a series of latticed ventilators. In building the side walls, every alternate ten feet, was left open from ground to roof. These open spaces were fitted with roller screens of jointed, wooden slats, operated by weights and springs, which allowed the interior to be well lighted and thoroughly ventilated. These screens could all be raised or lowered at pleasure. While the barns were being filled, they were all open.

"As the fields of grain commenced to ripeh'wue li

the straw was still green and full of sap, and the swollen kernels were just passing out of the dough stage of maturing; with the aid of a large force of workers, operating improved machinery, entire fields of standing grain at just precisely the proper stage of maturity, could be transferred to the shelter of these barns in a single day. As the heavy green bundles of grain were delivered from the fields, to the adjustable elevators working through the open spaces of the barns, from either side, these bundles were carried to the hands of the rick-builders, who piled them into narrow ricks five feet in width, across the barn and up to the roof. As the ricks grew in height, strong wire screens were hooked to the dividing posts which marked the boundaries of the ricks. These screens kept the bundles in place, and the ricks securely upright. When the barns were filled in this way, the ricks were separated by four feet of open space, with a ventilator in the roof for each pair of ricks and spaces.

"When the grain crops were thus housed without waste from shelling, the curing process went forward swiftly and securely. The advantages gained, were many. The wheat straw, full of sap when harvested, in curing slowly, kept the plump kernels of grain from shrinking, while it left them with clear, smooth, thin skins, and a quality, which produced less bran and more gluten, in the flour they would yield when ground. The kernels were all more uniform in size, larger, firmer and fairer; would all grade as number one. No sprouted wheat! No must! No blight! No rust!

"This was also true of oats and barley. The straw came from the improved threshers, in straight, compact bundles, thoroughly freed from grain, fragrant and bright, almost as nutritious for forage as hay. In fact, this straw, in such excellent shape for cutting, feeding, storing, or transportation, possessed more than twice the selling value of the best of ordinary straw. The oat straw, being softer and more pliable, was still more valuable as forage. The barley straw, less desirable for stock food, was sent to the paper mill for the use of the box factory. By this method of harvesting and curing grain, the increase in quality and selling value, was largely augmented. The general result was a marked saving of grain, time, labor and money.

'In cutting and curing the hay crops, the same kind of barns were used. The loosely packed hay in the tall, thin ricks, was soon dry enough to bale, and then be transferred to the storing barns ; leaving room for the corn crop which was to follow. Hay cured in this way is superior to anything on the market, and always brings tip-top prices !

"In curing corn, more time and wider ricks are necessary. The corn could be cut earlier, thus leaving the ground free to be prepared for the succeeding crop of fall wheat or late vegetables. During stormy weather, after this slower curing process was complete, a jolly army of huskers invaded the barns. The ripe corn, free from husk, was carefully assorted and stored in the ventilated bins prepared for it. The selected husks were packed and baled, ready for market. The stalks were stripped and topped by a clever machine. The excellent forage thus accumulated, was baled and stored. The pith in the large part of the stalk, was then extracted by another machine. These piths were then treated to a water-proofing process, sent to a shop on the farm, and made up into life preservers. Both life preservers and life rafts, made from pith treated in this way, proved lighter, cheaper, and more buoyant than those made from cork. This, you will observe is another profitable industry, added to the financial resources of Solaris. It is also an addition to the fitting employments for women.

"A still more desirable employment for our women co-operators, was found at the grain mill, where wheat, oats, and barley were transformed into popular brands of 'Solaris Breakfast Food.' Thus prepared, the market value of a bushel of grain was increased four fold.

"A new food preparation, from a mixture of popcorn with equal parts of thoroughly ground, roasted sweet corn, is really an excellent article of diet. In small, neat packages, this healthy and attractive food can be sold at a large profit.

"All of these sources of profit, naturally grow out of the new methods of harvesting and housing grain, which is made possible by the curing barns. While in appearance, these barns may not prove attractive, yet, I think you will readily acknowledge that they are very useful buildings; buildings which Solaris could not well do without."

"Really! Fillmore, I think these buildings are very fine! More than that, they are wonderfully well adapted to the purpose for which they were constructed! In this respect they certainly excel in usefulness, all other classes of barns. In your discription of them, and of the new methods in harvesting; I have been as much interested and entertained as though you were relating some fascinating romance. Indeed, I have been so absorbed, that I fear my poor note-book has been sadly neglected!

"How much land do you devote to cotton growing?

How has co-operative methods, affected its culture as a paying crop?"

"Last year, we planted twelve hundred acres in cotton. By the use of choice seed, a weedless soil, improved methods in the destruction of insect enemies, a better selection of fibre-producing fertilizers, a less wasteful plan of planting, and a more careful culture, we have increased the yield per acre from 300 to 500, and in a few instances to 550 pounds. When the crop was picked and ginned, we had twelve hundred bales of fine cotton. The quality of the fibre in the whole lot, was so excellent and so uniformly well ripened, that we were offered two cents per pound above the ruling price of ordinary cotton. As a result, this one crop gave the farm a cash income of $65,000. $60,000 for the fibre, and $5,000 for the seed, oil and oil cake. Choice seed for planting, was a large item in the last named amount.

"Heretofore, the great difficulty experienced by single farmers in growing large crops of cotton, has arisen from the want of sufficient help during the picking season. At Solaris, we always have an abundance of help. If the needs of the work seem to demand it, we can put two six-hour reliefs of pickers into the field each day, with 200 pickers in each relief. By working such a force, a large crop can soon be gathered without waste or damage. The pickers, all receiving the same daily wages, have a pocket interest in saving the cotton, therefore clean, careful picking, with a view of preserving a high grade of fibre, soon becomes the rule. This is an important matter, as green, immature fibre is worthless for the purpose of making a strong, durable thread or fabric; therefore pickers must be sufficiently intelligent, to understand why they should select only the thoroughly ripened cotton.

"Care is taken to make the pickers as comfortable as possible. For this purpose, broad, movable awnings, are provided to protect them from sun and showers. Under such circumstances, the picking season becomes one of fun and frolic, to which our co-operators, look forward with rejoicing. Six hours in each day spent in such light, pleasant work, is hardly regarded as toil. Yet, the amount of cotton picked by each individual, measured by the number of hours employed, is fully up to the standard set by good pickers, under the old system of long hours. The nimble-fingered women easily bear off the palm, as the expert pickers. If they were paid by the pound, their earnings would be greater than those of the men. Judged by such practical work, women cannot much longer be classed with the weaker units of an agricultural colony!"

"I consider that, as a very important point, well stated! But pardon me Fillmore, for the question! You spoke of better methods for the destruction of insect enemies. What are those insects, and how did you manage to destroy them?"

"Those that proved the most troublesome, were the cut-worm and boll-worm. Both were hatched from the eggs laid by certain kinds of moths. During the nights of the egg-laying season, for these moths, they were easily trapped and destroyed. By the use of a large number of electric light traps, suspended from convenient wires, thousands of these insects were lured to destruction before they could deposit their eggs. We are encouraged to believe, that a few years of such wholesale extermination, will soon rid us of these pests altogether.

"With a view of securing a continuous improvement in the quality of the cotton, we propose during the next five years, to carefully select the seed for each successive planting, from the largest, most prolific stalks, that produce the finest fibre. Reasoning from past experience, I think it will not be difficult to obtain a yield at least one-third greater than that of last year; which, on account of extra-superior quality, will readily sell for a still higher price. A careful reading of the annual reports, made by our consuls, who are stationed at the principal commercial ports of the world, has taught us, that to sell well, American cotton must be baled to meet the requirements of foreign markets. These markets demand that we must use a finer, better quality of baling burlaps, that will enable us to make closer, stronger, smoother packages, such as will at once impress the prospective buyer with the fact that they are really fine, because in appearance they are so tight, tidy, and attractive. To secure this, a small additional expense for baling material, is money well spent.

"Considering cotton as a cash crop, our experience so far, proves it to be especially adapted to the needs and methods of co-operative farming. A single crop has put money enough into our treasury, to pay more than double the purchase price of this farm."

"From your very clear and comprehensive answers to my questions, it appears that a co-operative farm, by reason of the number and organization of its workers, is equipped to carry on the culture of cotton with more than ordinary profit. This I accept as being absolutely true! Therefore I hail your success as a revelation of new possibilities, which must surely follow in the near future!"

CHAPTER XXXI.

THE DISCUSSION GROWS MORE INTERESTING.

"NOW FILLMORE,"said Fern, "I wish to ask, what have you been doing in the department of experimental farming?"

"Much of the work in that department is still in such a preliminary stage, that definite results cannot yet be declared. However, among the experiments worthy of mention, are the fields containing the various kinds of true sugar cane, and of sorghum or Chinese sugar cane.

"By hybridizing and other methods, we are striving to increase the hardiness of the former and the crystallizing-sugar product of the latter. By the results already obtained we are encouraged to believe, that five years hence, we shall have produced a sugar-cane equal to the best, that may be grown with much profit, as far north as St. Louis.

"Small plots of ground have also been devoted to growing tea, peppers, sage, hops, ginseng and other medicinal plants, with such excellent results, that no doubt they will soon develop into profitable ventures.

"The ten acres planted to broom-corn, have produced the necessary material with which to keep the workers in the broom and brush factory profitably employed.

"In the line of fibre plants, other than the cotton crop before mentioned; we have grown enough hemp and flax, to supply the needs of our rope and twine works. In 'bromelia fibrista,' a new fibre plant, we find a product that bids fair to rival silk in producing a fabric of fine, smooth, beautiful texture.

"In addition to the foregoing, several swampy plots have been planted to willow, and as a consequence, a growing basket-weaving industry has been developed.

"At the very beginning of our work here, while I was preparing to stock the seed beds in the nursery, one of our co-operators, a very intelligent and observing young man, who had been railroading in Mexico for two years previous to his joining our colony, called my attention to the Mexican quince. So strongly did he assert his belief that the fruit would thrive at Solaris, that I soon became a convert to his enthusiasm. With the young man for a guide, two weeks later we were on the way to Mexico; returning shortly, with enough three-year-old nursery stock, to plant one hundred acres. In addition, we secured the seed for 500,000 young plants. Since that time, our plantation of quince bushes has grown finely.

"Last year we gathered the first crop. Not a large one—perhaps, from fifteen to twenty-five quinces from each clump of bushes. As the fruit was large and the bushes thickly planted, the yield was about one hundred crates to the acre. An aggregate of ten thousand crates for the entire crop. We have every reason to believe, that the crop this year will be double that amount.

"Owing to the fact that this quince thrives best on the elevated table lands of Mexico, where it is subject to periods of cold and frost of considerable length; it has readily adjusted itself to this location and climate. We are now able to pronounce it, a complete success! It is a magnificent fruit!

Much superior in size, color, flavor and fragrance, to our own domestic quince. In keeping qualities and a firmness of flesh that will bear long distance transportation without injury, it is fully equal to the northern quince. In a deep-toned richness of color, perfection of shape and smoothness of skin, these peerless quinces are veritable apples of gold! They are pictures of beauty which sell at sight! The flavor is so fine, that Mexicans eat them with as much relish as the people of New York eat apples. Dried, these quinces are delicious!

"In Mexico, large quantities are annually reduced to a soft mass of pulp, spread out in thin layers, and dried into sheets of what is termed quince-leather. Armed with a generous roll of this excellent preparation, the traveler in the desert countries of hot, dry climates, may bid defiance to thirst. With such a wealth of recommendations, we were able to sell our first crop of quinces at a net price of two dollars per crate; or $20,000 in cash. Hereafter we shall save the commissions, as we have already received advance orders for our next crop, at $2.25 per crate, delivered on board the cars here at Solaris. Next year, we propose to enlarge our quince orchard by adding another hundred acres. Taking all these items into consideration, I think we have good reason to be proud of our first attempt at experimental farming in the line of quince culture!

"I have two additional experiments to describe. They are the last on my list.

"While in Mexico securing the quince plants, I found what to me was a new variety of table grapes. They were marked by the following characteristics. Large clusters, berry large oblong, thin skin, few seeds, fine sweet pulp, delicious boquet, color when ripe, a pale amber green; ripens about the first of July. As we found these grapes growing on the high table lands, I determined to try them at Solaris. By the dint of hard work, I procured enough young vines to set fifty acres. From those vines, we have rooted enough cuttings in the nursery, to give us 100,000 young vines, which have now reached the proper size for setting in the vineyard. This fine grape we have named 'Solaris Early.'

"Last July we gathered our first crop—5000 ten-pound baskets, which we readily sold at the fancy wholesale price of one dollar per basket. In packing them for the market we carefully reject small, poor bunches. The bunches selected are freed from all bruised berries. The stems of the bunches are then dipped in melted wax. After this treatment they are packed in layers of finely cut, soft chaff, made from clean, bright, fragrant oat straw. The chaff serves to keep the berries and clusters well apart, and also to keep out the air, which otherwise would soon wilt the fruit. Packed in this way the grapes reach distant markets in perfect condition. In fact, they are the only good table grapes on the market at that season; therefore in choice lots they will always command fancy prices. The experiment with them has proved so successful that next season, we shall increase the size of the vineyard to two hundred acres.

"By way of a commencement in small fruit culture, we have fifty acres of ground, devoted to growing a great variety of berries. They require the work of a large number of hands during the picking season. Owing to the perishable nature of such small fruits, we do not attempt to market them fresh, but make them into jellies, jams, marmalades, and preserves.

These we pack in glass jars, of the various sizes demanded by the wholesale and retail trade. In preparing and packing these goods, we use only the best of everything. This is in line with our purpose to establish a reputation of a high degree of excellence, for each article put on the market under a Solaris label. By a rigid observance of this rule, we manage to sell the products of our berry crops at a good profit.

"When the farm books are balanced at the end of the year, we are encouraged to find that the fifty acres of berries, has a larger credit than any other fifty acres on the farm.

"In the line of an extension of this kind of farming, we are now preparing for next year, with the purpose of starting a factory for canning our output of sweet corn, green peas, beans, asparagus, tomatoes, peaches, plums and pears. This completes my list of items under the head of experimental farming, which Solaris now has to offer. What do you think of it so far?"

"I think very well of it indeed! I am especially impressed with the Mexican quinces, early grapes, and the berries. They seem to promise the greatest success, and the largest financial returns. Taken altogether, I think the outlook for experimental farming at Solaris, is very bright!

"Now, by the way of recapitulation, can you give to me, a brief statement of the crops grown last year; with an approximate one, of the cash derived therefrom?"

"That will not be difficult. I will endeavor to make my statement as brief as possible.

"By looking at this map, you will observe that during the season just past, we have cultivated about 4,000 acres of land. The crops planted, were nearly as follows: 1,200 acres to cotton; 1,000 acres to wheat; 1,100 acres divided between corn, oats, barley and hay; 150 acres to vegetables, and 550 acres to a miscellaneous variety of crops, such as the nursery, the quince orchard, the vineyard, the berries, the gardens, and all ground devoted to experimental culture.

"The aggregate cash income derived from these crops, which found a market in the outside world, in addition to those sold to our own people, amounted in round numbers to $193,000. Of this amount, $95,000 came from sales of cotton and wheat. Next year we have good reason to expect a cash income of $250,000 from our farm products alone. Last year we realized $57,000 from the sale of our manufactured products; such as brick, terracotta, drain pipes, tiles, earthen ware, furniture, brooms, willow ware, and the output of several other minor industries. This brought the total income of the farm for the year, up to $250,000.

"You ask what disposition has been made of this money? $50,000 has been expended in additional improvements, machinery, buildings, and live stock for the farm. $25,000 more, has been added to the stock in our store, which now has a supply of goods, sufficient to meet the demands of adjacent settlers who wish to trade with us. $25,000 is held in our treasury, for use in any emergency which may arise. The remaining $150,000, has been placed in the sinking-fund.

"Our farm-store, has proved a very important institution. The clothing, tailoring, dressmaking and millinery departments, have proved surprisingly successful; with a constantly increasing demand for the goods turned out. This opens a wide field of remunerative labor, for our women co-operators.

"The 2,400 acres of untilled lands, are now utilized as follows: 500 acres are covered by a fairly good native forest; 500 more, by the scattered timber around the stone quarries, gravel beds, sand pits, clay deposits and the various other mines. 400 acres are used for pasture, 100 acres belong to the village site. 200 acres are planted to apple trees; 25 acres to pear; 25 acres to peach; and 200 acres to nut-bearing trees. 100 acres are now being prepared for the addition to the quince orchard. Another 100 acres for the vineyard. The remaining 250 acres, for other desirable varieties of fruit.

"Of the 100 acres set apart for the village site, only forty, are at present occupied by the streets in use, the buildings, and the public square. The remaining sixty acres, are laid out with walks, drives, lawns, oval, circular, and star-shaped plots. The latter, are filled with choice roses and flowers. The ovals and circles, are thickly planted with fruit trees and ornamental shrubbery. The fruits, such as cherries, plums, peaches, pears and figs, have all been the result of experimental potting and planting by the school children. The same is true in a large measure, of the rose gardens and the shrubbery.

"The effect of this amusing work on the children, is most excellent. A taste for the beautiful becomes permanent, while they acquire a fund of useful knowledge about the care and culture of trees, and also how to enjoy themselves in the conscious zeal of pushing forward some useful employment; which will make them stronger, healthier and happier. With the advent of spring, comes a wealth of bloom to reward their toil—a paradise of beauty and fragrance; everywhere, clouds of pink sprays and snowy petals charm the sight.

"This last item, like a long, ornamental flourish, must conclude my summing up of the distribution of crops, the division of forest, pasture and fruit lands, over the whole farm; with its complete chain of financial resources, and its outlook for the coming season. I hope I have not made my recapitulation too lengthy! Also, that I have succeeded in answering your questions satisfactorily."

"Your summing up has shown surprising results! The magnitude of the cash income, is really a crown of triumph for co-operative farming! I congratulate you, and the people of Solaris, most heartily! In justice to the able answers to my questions, I must say that many times you have answered, even before I could frame them into words. With each succeeding reply, my wonder and delight has increased. I have discovered many new possibilities, in pleasant, productive and profitable methods for farm work, of which I have never before dreamed. Now that you have made them plain to me in such a charming manner; I am beginning to understand how it is, that Solaris can produce such quantities of marketable goods, that can so easily be turned into cash. I have yet a number of important questions remaining unanswered, but they do not pertain to growing crops."

CHAPTER XXXII.

SOCIAL SOLUTIONS.

"I NOW wish," said Fern, "to consider the social and domestic interests of the colony. How do you manage to keep up the necessary degree of cleanliness, demanded by perfect sanitation in the living rooms of the co-operators, without seriously disturbing the privacy of the family."

"That is a delicate matter, which by choice of the co-operators themselves, easily adjusts itself to the requirements of the committee members, who are chosen to take charge of the tri-weekly scrubbing and sweeping. The detail for this work for each week, is made by the assignment committee.

"They select from a class of workers, known as both skillful and trustworthy. All rooms which the occupants desire to have cleaned, are left open. All rooms that are found locked, are reported to the chairman of the committee, whose duty it is to inspect them at a later period, while the occupants are present. It is a matter which is well understood by the members of the company, that rooms not accessible to the regular cleaning force, must be kept sweet and tidy by the occupants themselves, during hours which might be otherwise devoted to rest, amusement or study.

"Under the pressure of such conditions, even the most exclusive, soon voluntarily open all their rooms to the authorized force. Causes for complaint against any member of the sanitary, inspection or assignment committee, are corrected by the voters at monthly elections, held for the purpose of selecting new committees. This system so appeals to that innate sense of justice and harmony reigning in the hearts of our people, that after a few months of experience, they are ready to co-operate heartily in any sort of discipline which may be necessary to secure the welfare of the entire colony.

"The peculiar charm of colony-life appeals to them so strongly, that to be voted out of the organization on account of violation of rules, or of any improper conduct, is universally considered as a most dreadful calamity. The possibility of such a fate, like some hidden spectre, acts as a restraining influence, which holds in check the most lawless, stubborn, or self-opinionated. It soon makes them zealous, peace-loving and obedient. Having once tasted the sweets of the co-operative system, they have a wholesome dread of being obliged to return to the cruel bitterness of the old competitive system!

"Among the most potent charms which have proved so attractive to Solaris workers, is the condition of health, comfort and beauty, which surrounds the laborer in every department of the farm.

"In store, work-shop, seed-room, dairy, mill, factory or packing-house, the rooms are large, the light is abundant, ventilation perfect, ceilings high; while both walls and ceilings are so beautifully and artistically decorated, that love for the beautiful in the esthetic nature, swells and grows to be a dominant passion. This passion soon takes hold of both heart and brain, becoming the foundation of a character-building-work of high order. Thus happily environed, our people feast their eyes and merrily sing away the hours, which are devoted to tasks they have learned to love. The tendency of these things, is ever toward the good, the right, the pure and true! Under such conditions, the demon of discontent, evil thinking and evil doing, cannot thrive! His power wanes, he flies to the more congenial surroundings which mark the dingy, ill smelling, overcrowded work-shops of the competitive system!

"No wonder, when away from Solaris, our people are so anxious to return! They come back convinced, that they have fortunately escaped from the thralldom of a debasing, cruel system. A system which—utterly ignoring the sacredness of human life—in a frenzy of selfish greed, has, so far as the toilers of the world are concerned, turned the triumphs of modern civilization into the mockery of a bitter curse! As affecting themselves, our people perceive that, under the protecting mantle of financial conditions which prevail here at Solaris, they, as members of the company, are sure to secure every benefit, profit or advantage, that may flow from the use of the best and most expensive kinds of labor-saving machinery. Once aware of all the facts, thereafter, they cannot under any circumstances, be induced to return to employment under the old system.

"The advantage in favor of co-operative work is so great, that among our women co-operators, there is a general desire to have it utilized to the utmost; especially in all kinds of housework. The introduction of such a wholesale system of house-cleaning, soon demands a better class of sweepers, to take the place of the housewife's broom and dust pan.

"Large suction sweepers, worked by a powerful inhaling bellows, which swiftly and silently suck up, from carpet, furniture, and curtains, all particles of accumulated dust, are the perfected instruments chosen; unlike the ordinary dust-raising machines, which must be followed by an army of dusting cloths, these suction machines do perfect work, leaving the air of the renovated room pure, wholesome and fairly free from floating dust, with its accompanying cloud of disease-laden germs. Many similar accomplishments in other departments of housework, soon convince all opponents, that personal prejudice must not be allowed to interfere with the working of the system."

"Pardon me Fillmore! If at this point I interrupt you, with a question which I wish to preface with this remark! In the estimation of most women, well-kept hands, are considered as a rule, to indicate the measure of the owners refinement. According to my judgment, there is nothing which so quickly destroys the contour and suppleness of the hands, and that much prized, white, velvety smoothness of skin, as dishwashing. As a matter of fact, the woman's self-respect is involved in the loss. For this reason, I believe women dislike that disagreeable part of housework more than any other. Premising that my theory is true, how can you manage this matter at Solaris, in order to avoid trouble?"

"I accept your question as a welcome interruption! It gives me a chance to tell you more about our kitchen work, which I feel sure will interest you greatly!

"For reasons which I shall state presently, our women workers do not desire to avoid frequent six-hour details as dishwashers at the restaurant. By our new methods, the task is easily and quickly accomplished.

"The washers are not required to put their hands into hot or cold water during the process. Traveling carriers on either side of the dining rooms, run to and from the kitchen. In one, the food comes to the tables, in response to phone orders from the waiter. In the other, the dishes are returned to the kitchen. There, the washers scrape the bones and rejected food into the waiting barrels. These barrels when filled, go to the feeding yards of the pigs and poultry.

"The dishes, after being scraped, are then placed in the washing machine. This machine, run by electric power, is a wide, deep, round-bottomed trough, built in a circle twenty feet in diameter. Along the bottom of this trough, is a moving track, which travels slowly around the circle with its train of metal carriers. On these carriers are placed the dishes as they come from the hands of the scrapers. When the carrier thus laden commences its circular journey, the dishes—placed well apart—are subjected to dashing jets of warm, soapy water, and then to more torrential jets of hot, and very hot pure water.

"Comfortably seated, at convenient points around the machine, the washers control the force and quantity of the water jets, and whenever necessary, assist the cleansing process with their long-handled swabs. When this process is finished, the dishes arrive at the drying boards, so hot that by the time the wipers with their thick towels have placed them in the racks where they belong, all are perfectly clean and dry.

"Our pots, sauce pans, stew pans and kettles, are all designed for electric cooking, and are made in shapes best adapted for easy cleaning. For these, an additional washing-sink is provided. Over this sink, connected with the electric wires, we have rigged three hanging spindles, of as many different sizes. These spindles can be raised or lowered by the operator, while they are in motion. Each spindle is armed on every side with loose wings of alternating wire scrapers and dish-cloths. The vessel to be cleansed is placed on the movable carrier at the bottom of the sink. Passing under a spindle of the proper size, the spindle is lowered, and at once begins to revolve with a strong, rotary pressure. This searching, chafing pressure, in connection with the hot-water jets, soon cleans and polishes the most obstinate among the kettles.

"The kitchen and dish pantry combined, is a very large, well-lighted, well-ventilated room. This room is constantly kept sweet and comfortable by electric fans. The work is light, and never monotonous. Only two, of the six hours devoted to kitchen duty, are spent in the active work of dish washing. During the remaining hours, the washers take lessons in cookery, from the chief and the two assistants. These three important officials, are chosen from the ranks of competent volunteers. They are responsible for the kitchen work. They plan all the meals, and direct the work of the under cooks. The system soon comes to work like a charm! I can truthfully say, that it gives general satisfaction.

"The success attending this extension of co-operative methods, to embrace the entire list of worry-producing details which belong to general house work, is hailed with delight by our matrons and maidens. They keenly appreciate the great blessing of this movement, which has rescued them from the harrassing, health-destroying drudgery, of a house wife on a small farm. They well know the sad story, which comes from thousands of such farms, where isolated lives, overburden of cares and long hours of irritating, never-ending toil, have produced such fearful, mental depression, that as a result, we find six hundred farmers' wives, among the inmates of asylums for the insane, in each one of the States of Michigan and Kansas. The proportion for other agricultural States, is doubtless much the same. What a horrible array of statistics, this is to contemplate! What an indictment against existing agricultural conditions! What a sad fate, to overtake the mothers of so many sons and daughters of the farms of this Republic! Who can measure the intensity of the agony and suffering, these children may thus inherit! What possible argument, can speak more eloquently, or call more loudly, for the immediate adoption of co-operative farming by our agricultural people?

"In the matter of frequent bathing to maintain personal cleanliness; the popularity, with both old and young, of our fine hot and cold, plunge, swimming and shower baths, free to all, which are kept open in connection with the laundry; proves conclusively, that the habit of cleanliness, like all other habits, is the result of environment; or in other words, of opportunity and the strong impulse of social example.

"In treating your question as though it contained several sub-divisions, I may perhaps have made my answer too lengthy. Do you find it so?"

"Oh no! On the contrary it is clear, brief, interesting and to the point! You have told me just what I most desired to know! I perceive that the practical working of a co-operative colony, answers a great many puzzling questions, which hitherto, we have passed by as hopeless problems. From the commencement of this work, I have been concerned, lest the discipline necessary to maintain a proper working harmony in such a large colony, should prove a fruitful source of discontent. I am rejoiced to find that my fears were groundless!

"This brings me to my second question. Do you find homesickness among the colonists, a frequent cause of discontent?"

"On the contrary, the number of such cases has been surprisingly small. Owing, doubtless, to the marked change from isolated conditions of small farm life, to the superior advantages for education, amusement, social enjoyment, and the all-pervading enthusiasm of congenial, co-operative work; which here at Solaris, leaves no time for such fits of brooding over the past, as usually result in that severe mental depression, which we call homesickness. Perhaps one individual in fifty, is so constituted that home sickness becomes a serious illness. In such cases, the executive committee is authorized to grant the necessary leave of absence. Always providing of course, that the applicant is willing to comply with a rule of the organization, which assigns the pay of the absentee to the general service fund, for the number of days such absence may continue. A strict observance of this rule, leaves no cause for complaint by those who remain.

"In considering the question from another standpoint, we find the general tone and disposition of our people, has been raised to a much higher, happier pitch, by the evolution of the musical spirit, introduced and inspired by the work of the dancing and musical clubs. Stimulated by the prizes offered by the general manager, a great number of beautiful farm songs have been completed, and adapted to a large variety of farm work. These songs have been taken up by a goodly number of glee clubs, organized for the purpose from among those members of the musical club, who had the good fortune to possess a fine quality of voice.

"Careful training and steady practice, soon enabled these lesser vocal organizations, to render the entire list of songs, with a mellow smoothness, an inspiring swing of rythm, and a well rounded tone of perfection, which was really quite surprising. These vocalists, scattered through the fifties and hundreds of farm workers in the hay, harvest, corn and cotton fields; the nursery, gardens, orchards and vineyards; the dairy, mills, factories and packing-houses; the brick works, mines and quarries; the workshops of the store, and the assembly meetings of the co-operators; became competent teachers, who, by their leadership and example, soon made it possible for every member of the colony, to master both words and music of all the songs This course of vocal training proved so fascinating, that our people literally absorbed it! The children, even more quickly than the adults!

"Thoroughly tested in the practical work of every department of the farm; the beneficial effect has proved a marvel, which has far exceeded the expectations of our musical enthusiasts. Many fine voices have been discovered, developed and trained. The benign influence of this musical wave, has shown a constant tendency to extend its sway in all directions. This blending of voices, has added a hitherto unknown zest to the work; and a stronger tie to every association connected with it. Best of all, as directly affecting the question under discussion! It has proved a most potent factor in driving away the spirit of ill-humor, inharmony, and discontent; also in breaking the charm of old associations, home ties, and retrospective, social memories, so conducive to attacks of homesickness. The exhilarating, helpful rythm, of these inspiring songs, has given an added force to the working power of the farm. It has largely reduced the fatigue, and increased the amount of work that can be performed in a given time. Further, we find the general mental, physical and spiritual health of our people, correspondingly improved.

"A curious fact, is disclosed by these vocal experiments. It is this, that the vibration of musical tones, in the blending voices of a mixed multitude, produces a moral, mental and spiritual harmony, such as cannot be achieved in any other way. In point of fact, we get a composite expression of the highest soul element of the mass—a new phase of the exceeding fruitfulness of co-operative effort! It may be stated in conclusion, that there comes to the minds of our people, an added power, flowing from the general hypnotic effect, of harmonious co-operation. This power brings with it a right conception of human life, in which a certain amount of necessary, productive labor, becomes the keynote, which completes a perfect anthem, and more symmetrically rounds out the full measure, melody and grandeur, of an individual existence. What think you of these results?"

"They are very wonderful indeed! They reflect much credit on the excellent work inspired by the dancing and musical clubs; also on the genius and culture of the vocalists, and the marvelous efficiency of a well-directed co-operative effort. This triumph in a new field, which so increases the possibilities of soul expression, suggests the use of music as a prime factor in all future systems for ethical culture.

"Now Fillmore, please tell me. How has the example of Solaris farm, affected the industrial, social, and political situation in this town and county?"

"The effect has been favorable in every way! The attractiveness of our social organization! the financial success which has crowned our farming and manufacturing operations; the opportunities offered for young men to learn so much of the industrial arts; the short hours of light labor; the long hours of leisure for rest, study and amusement; the educational, health-giving character, of the amusements; the fascination, of the club-system of education for adults; the irresistible charm, of the dancing and vocal entertainments; the generous wages paid to the co-operators, which affords for them such an abundant supply of food, clothing and books; the fine quality and perfect reliability of the large assortment of goods in the farm-store; the advantages of a rational scheme of insurance, which stands as an absolute safe-guard against accidents, sickness and old-age; the improved conditions for women, which largely relieves them from the irritating, nerve-destroying worry, of a constant burden of household cares; the fostering care for children, which insures for them ideal opportunities for birth, unfoldment and education; the manifest advantage of farming on a scale large enough to allow the use of the latest and best labor-saving machinery; the astonishing array of huge, modern barns, storing, curing and packing houses; the wonderful cheapness and utility of the electric power; the long list of farm implements, many of them especially invented, which followed the introduction of this magic-working power; the wide publicity given to these things through the columns of the Solaris Sentinel, our weekly farm paper, sent free to friends of the colonists, and to all who ask for it; considered altogether as a comprehensive whole, is a startling combination, which has arrested the attention, aroused the interest and provoked the astonishment of surrounding communities, far and near. As a consequence, our office has been overwhelmed with a flood of correspondence from interested enquirers, followed by an ever-increasing stream of visitors to Solaris, to see for themselves, the verity of this twentieth century model of farm innovation. In order to answer the great bulk of queries, emanating from these two sources, a series of articles describing the object and purpose, and explaining the details of the enterprise, has been prepared for the columns of the Sentinel. With an extra large edition of this newspaper, we are prepared to supply as many interested people as may apply.

"The applications to join the company, made by progressive young farmers in this and adjacent counties, have become so frequent and persistent, that finally we have consented to prepare the leaders for another co-operative colony, which we propose to locate on a certain one, of the nine remaining Fenwick-farm-sites, which happens to be in this county, only ten miles distant from Solaris. This preparatory class, is limited to fifty people; one-half females, married couples ranging from eighteen to twenty-five years of age, preferred. The course for this class, contemplates one year of practical work, embracing all departments of the farm.

"The membership of this class, was filled six months ago. Six months hence, the graduates will be prepared to organize the new colony. I am greatly interested in the scheme, and have promised to aid in every possible way.

"To this body of pupils, is referred all applications from prospective co-operators. Judging from the mass of applications already accumulated, when the time of organization for the new colony arrives, the list of eligible applicants will probably contain a thousand names. The outlook for the new farm company, seems unusually bright!

"Both board and tuition for these pupils, are donated by Solaris Farm. At the end of the year, $100 in Solaris scrip, will be paid to each one, as some sort of compensation for the year's work. This arrangement is accepted by the pupils, as fair and perfectly satisfactory.

"Referring to the relations existing between the Solaris Farm Company, and the township and county officials. It is noteworthy, that no serious friction has arisen. One year ago, a large proportion of town officers, including the assessor, town clerk, magistrate and chairman of the Board of Supervisors, were chosen from Solaris. Owing to the small, much-scattered, population of this county, the present county sheriff, auditor and treasurer, are also Solaris co-operators. The manifest integrity of this institution, seems to be accepted by the voters of the county, as a guarantee of the honesty and ability of its members. The significance of this approval, so early in the history of the movement, augurs well for the future dominancy of our social and industrial system, as a political factor in both town and county.

"The Solaris Company has erected a roomy, substantial building, for the use of the town officials, for which a moderate rent is paid from the town-treasury. The county officers have secured one hundred acres of land two miles from Solaris, just outside the farm limits. On this, they propose to erect a suitable brick building for the county offices. The farm company, now has the contract to furnish the brick and erect the building. Pending its completion, the county officials occupy rented quarters in Solaris, which is by far the largest business center in the county. From this statement of the situation, you will observe that our co-operative vote already holds a balance of power, which controls the policy of both town and county. With the advent of Colony number 2, the interests of co-operation in this county, are secure for all time. Meanwhile, we are encouraged to hope that before the close of the twentieth century, what co-operation has already achieved at Solaris, may be accomplished in every town, county and state in the Republic!

"You ask, what disposition is made of the salaries of such co-operators as are elected to fill town and county offices?

"They are paid in scrip. The salaries or fees which they receive from town or county, are turned into the company treasury. As these co-operators, in holding such offices, are in a position to materially aid the co-operative movement. They are justly excused from farm-work, whenever their official duties require attention."

"Splendid! my dear Fillmore! Your report is very interesting, and even more encouraging! It seems the beginning of a fulfillment of my father's hopes, dreams and prophecies! I am anxious for the time to come, when he can tell you how much he is pleased with your work!"

CHAPTER XXXIII.

SOLARIS SCRIP.

"RETURNING again, Fillmore, to the financial operations of the farm; with such a volume of business to transact, how do you manage to get along without having recourse to some local bank?"

"To a large extent, we do our own banking business. Our treasurer, has his office in the cash room of the store. In this room we have a large vault, containing a fire-proof safe of the latest type. The books, records and funds of the company, are all kept in this safe. For our commercial business, we have selected one of the principal banks of St. Louis as our bank of deposit. A large percentage of purchases for the store and farm are made in that city, which is also a market for the bulk of our farm produce.

"The farm company has an office near the bank, where some member of the executive committee, or other representative of the company, may be found every business day of the year. It is the duty of this agent to attend to purchases, consignments and sales; also to have charge of all business transacted through the bank of deposit. Taking care, to keep the amount of available funds up to the ten thousand dollar mark. To do this, it sometimes becomes necessary for the company to issue drafts on the bank of deposit for thirty, sixty and ninety days. These drafts are accepted by dealers, for purchases made in Chicago, Cincinnati, Philadelphia or New York, the same as cash.

"As borrowers, our only dealings have been with

you. In these dealings, at times when much in need of more capital, we have not been required to pay interest. Now, having returned our borrowed capital, and being free from debt, we have grown more independent and self-sustaining; therefore more averse to the idea of paying interest to any one. We are convinced by past experience, that all necessity for incurring interest-bearing obligations can be avoided. The use of Solaris Scrip in all intercolonial transactions, has proved a most potent factor in helping us to arrive at such a fortunate conclusion. By its use, ninety per cent of our business can be transacted on a cash basis, without using one cent of actual cash. In addition, we can use it as a basis on which to borrow. To illustrate! Suppose we need ten thousand dollars to replenish the stock of goods in the store, pending the sale of products on hand. We borrow that amount from the insurance fund, the sum being part of the accumulated profits on sales at the store and restaurant. We then replace this sum by scrip of the same face value. This scrip, to the pensioner or beneficiaries, is the same as cash. When they have drawn and spent it, the debt is cancelled. No interest is paid. The store and restaurant become the clearing house, through which these drafts against the resources of the farm are liquidated. In the same way, temporary loans can be made from other funds, whenever it is for the benefit of the united interests of the cooperators to do so.

"How is it possible, you ask, to keep perfect control of such a large issue of scrip, with a certainty that all in use is genuine?

"That is a matter which is easily regulated by our simple system of issue. In the first place, we print the scrip here at Solaris, from plates which, when not in use, are kept in the safe, in the custody of the treasurer. The five denominations issued, are as follows : five, two, and one dollar bills; which, together with the fifty and twenty-five-cent, fractional-currency scrip, make up the list. Every denomination has a numbered series, of ten thousand. Each series, with the stubs attached to the bills, is bound in book form. When issued, each stub remaining in the book, will show the date of issue, serial number, and amount of the issued bill. When cancelled, the bills are returned to the book, and again attached to the stub to which they belong. At any time, an examination of the books of issued and unissued scrip in the hands of the treasurer, will give the amount outstanding. The co-operators are requested to keep a record of the serial numbers of the scrip they hold or handle, and to report the loss or destruction of such as may happen. A history of the loss is attached to the stub, and the amount of the bill carried to the profit and loss account of the company.

"If the genuineness of any piece of scrip should be questioned, a comparison with the stub should show the same date, number, amount and serrated edges, made by the peculiar pattern of the perforator belonging to that series. If so, the bill must be genuine. As time passes, we are more than ever convinced of the wonderful advantage gained by the use of this scrip. Our people find it much lighter and more desirable to carry and use, than the same amount of gold or silver coin; therefore they frequently request to be allowed to exchange coin for scrip. In summing up my replies to your questions: it seems probable, from the constantly increasing volume of business, that the company will soon be obliged to take a charter that will authorize it to do a complete banking business."

CHAPTER XXXIV.

THE INSURANCE OFFERED BY CO-OPERATIVE FARMING.

"I NOTICE, Fillmore, that you mention the borrowing of ten thousand dollars from the insurance fund; the same being a part of the accumulated profits on the business of the store and restaurant. Tell me; how is it possible for so large a sum to be saved in such a short time?"

"A complete answer to your question, will bring up the whole subject of insurance; which presents some interesting problems. I will first try to give you the basis for such an amount of savings. The net per-deim pay of $2.50 for each adult member of the company, will give an annual income of a little more than $900. If we include an added pro rata for the children, each one will spend annually at least $450 with the store for goods; and $350 with the restaurant for food Our statistics show much larger sums; but these will do for an estimate. Taking these figures for a basis, we find that the annual sales made to our own people by the store and restaurant combined, reach the startling sum of $400,000. A net profit of five per cent on this amount, gives $20,000 each year to the insurance fund. At this rate, the profits for thirty months, reach the goodly sum of $50,000. To which we may add $2,-

500 more, as profits on sales to the amount of $50,000, made during that period by the store and restaurant, to people from surrounding communities. Altogether, we have a grand up-to-date total for the insurance fund of $52,500. These profits will continue to increase with larger sales to outside people ; also with the increased wages or incomes of the co-operators, as the products and profits of the farm continue to grow.

"Such favorable statistics are very encouraging. They demonstrate that only a five per cent profit will be needed, to meet all future demands against the insurance fund, even when the colony has its maximum number of children and superannuated co-operators. The remaining profits, which in some departments of the store are large, may wisely be devoted to educational and missionary work.

"From another point of view, this eloquent array of figures, has an additional value. They show conclusively, that the restaurant alone furnishes a home market annually for $175,000 worth of farm produce: beef, mutton, pork, lard, honey, syrup, milk, butter, cheese, eggs, poultry, vegetables, fruits and grains.

"If we consider the sales made by the store, we find after deducting the cost of raw material, that at least fifty per cent of the goods purchased by our people, are really the products of the skilled labor of the farm: such as crockery, furniture, willow ware, picture frames, brushes, clothing, underwear, bed furnishings, and goods from the tailoring, dressmaking and millinery departments. From this showing it will appear, that the store becomes a home market each year, for farm products to the amount of $112,500. To this, let us add the sums of sales through the restaurant, and those made through the markets of the outside world. Altogether, we have a grand total of $787,500 for the market value of farm products last year.

"Does this exhibit appeal to you as a reasonable basis for the accumulated savings named in your questions ?"

"I am sure the exhibit has astonished me greatly! Your figures and statements are both fascinating and convincing. They are all, most excellent arguments in favor of co-operative methods. I now perceive that even on the basis of present conditions, a five per cent profit turned into the insurance fund, at the end of the first ten years, will amount to the extraordinary sum of $200,000. With this magnificent fund, you can afford to extend the scope of your original plan! How will you dispose of it? At what age do you propose to retire the active workers ?"

"Yes, our original plans have been changed, and very much enlarged. The insurance fund has grown so rapidly, that it was deemed wise to expend a portion of it, in building a hospital for the accommodation of our farm people, and perhaps a few outside patients. Last year, a two-story and basement brick building, was erected just in the heart of our finest shrubbery dotted lawn, some distance from the public square. It is large enough for about one hundred patients. Viewed from any point, it presents a charming appearance. It is conceded by all to be the handsomest structure on the farm. Inside, with its polished floors, magnificent windows, large rooms, high, beautifully frescoed walls and ceilings, dainty couches, cozy chairs, and wide, breezy halls, with picture-laden walls; every condition is present to satisfy the highest ideal of sick-room comfort.

Brighter, sunnier, more health-inspiring rooms never soothed, charmed or healed a nerve shattered patient!

"Under the supervision of the sanitary committee, the hospital at present, is in charge of a young surgeon employed by the company. His services are utilized in teaching and preparing a class of trained nurses. He also teaches the members of the chemistry and physiology clubs, in their new study rooms at the hospital. At a later period this surgeon will be superseded by two of our own people. A young woman and a young man, both with some previous knowledge of pharmacy, who have been in charge of the drug department at the store; have recently developed a strong desire to take a thorough course of medicine and surgery at some leading school. Upon the recommendation of the general manager, approved by a unanimous vote of the co-operators, the expense of this schooling is to be taken from the insurance fund, with the understanding however, that after graduating, they are to relieve the company of the expense of a hired surgeon, by taking permanent charge of the hospital, or as our people have christened it, the 'Temple of health.'

"Relative to the question of retiring members of the company; much thought and discussion on the part of our officers and co-operators, has been required, to properly and wisely fix the age at which such retirement shall take place.

"Many important questions have been considered. Our present colony, as you know, is composed of young people, as a rule not yet thirty years of age. Individually they possess strong, disease-resisting, vital organizations, which have been reinforced by harmonious, mental and physical development. This immunity from disease to such a large extent, has been still further strengthened and fortified, by the beneficial effects of our organized sanitary, social and industrial methods. These methods have lifted the weary burden of toil from our people, and substituted therefor, a light exhilarating labor, simply healthful exercise. Under such favorable conditions, our workers ought to reach the age of fifty, with health and vigor still unimpared. For the reasons named, very few of our co-operators, outside the ranks of the mother's club, are at present entitled on account of either illness or accident, to draw their wages from the insurance fund. Fortunately, so far, not one has become permanently disabled! All things considered, it was not unexpected, when a final vote on the question was taken, that a majority was found to be in favor of fixing the age of retirement at fifty years.

"This decision will give the farm company, twenty years in which to prepare for the event. In the light of our past experience, no one doubts our ability to accumulate an adequate fund, with which to meet the additional drain upon it. This drain will prove a heavy one, as the retired pay of the co-operators, who have reached the age of fifty, has been fixed at two-thirds of their present pay, that is, fifty dollars per month or $600 per annum. Premising that the maximum number on the retired list at any one time will not exceed fifty; the total annual retired pay will then amount to $30,000.

"The following plan has been devised to meet this additional expenditure. It has been demonstrated conclusively, that five years hence, the income of the farm, will warrant the increase of the wages of each member of the company, to $1,500 per year. At least $1,200 of this amount, will be spent at the store or restaurant. We shall then have a new basis for calculating the five per cent profit for the insurance fund; that is, $600,000 annually, which will give $30,000 each year for the fund. Allowing that savings at the present rate, $20,000 per annum, for seven and one-half years, aggregating $150,000; will prove ample for incidental needs, until the time for the retirement of the first co-operator! We calculate that fifteen years of savings on the new basis, will give us twenty years hence, a fund of $450,000 to commence with.

"If practical experience should prove that larger savings are necessary; an additional two and one-half per cent profit, may be set aside for this fund, without seriously curtailing the sums devoted to educational and missionary purposes. This will surely cover all possible contingencies. More especially, as seven and one-half per cent of all retired pay, will come back to the fund as profits on purchases—active workers having taken the place of the retired members. Considering the generous annuity provided by this insurance, together with the fact that the wants of the pensioners will become fewer as age increases; doubtless, at the end of each year, many of them will turn back into the fund, considerable sums of unused pay.

"As another important factor, connected with the question of this kind of insurance, it should be well understood, that after reaching the age of retirement, our members do not cease to be valuable productive workers, either for the financial gain of the colony, or for the general welfare of the movement, which the colony represents. On the contrary, in many cases, their services are liable to become more valuable than ever before. Between the ages of fifty and sixty, they remain subject to assignments to serve on committees, to act as traveling agents for the company, to represent the company as lecturers and organizers, for the spread of the movement; to act as aids to the teachers in the schools and the numerous clubs. They are also eligible to election as town, county, state or United States officials. In committee work, connected with the store and the various factories, their riper judgment, based on many years of experience, would prove especially valuable: often by timely advice, they would be able to save for the company in one transaction, an amount in money more than equal to their entire wages for the year.

"In another way their services would prove equally advantageous. With such an increase of leisure, there would come to these retired co-operators, a desire, and the opportunity, to enter more actively into the practical work of the scientific clubs. If inclined, they could take up all kinds of scientific research; making themselves especially useful in the practical, productive and profitable work of the educational, microscopical, chemical and photographic clubs. Those who had a talent for invention, could then devote as much time, energy and thought to it, as they chose. To aid them, they would have the advantage of an acquired skill in the use of tools, and of all kinds of complicated machinery, which would be a part of the outfit belonging to the thoroughly equipped machine shop at their disposal. In the laboratory, they could find the books, maps, and drawings, necessary to bring them up to date in any line of invention which they might chose to enter.

"Taking these important factors into consideration, we discover that our co-operative inventor, would be armed to conquer his subject by a magnificent equipment, such as an ordinary inventor could not hope to command.

"So ably reinforced by the advantages enumerated, our corps of inventors, of both sexes, would be inspired by a labor of love. Unbiased by any selfish motives, they would be working for the farm and for humanity. With no cause to distrust their fellows, they could openly discuss their discoveries, without fear of having them stolen; consequently, they could have the willing assistance of all the inventive minds in the colony, in developing and perfecting their original inventions. This would be an experience utterly unheard of, in the annals of an industry based on the competitive system. It would be the beginning of co-operative invention as an art. It would mark another great step in harmonious, practical and profitable co-operative thinking, that would lead to discoveries of vast importance to the world; discoveries that could not be made in any other way. It is difficult for even the most enthusiastic optimist to imagine, what a revolution in the inventive world, will follow the introduction of such superior co-operative methods; or what wonders will be wrought by them, before the close of the first half of the twentieth century!

"Let us consider what they might do for our superannuated farmers. Quickened by such an added potency of perfect, co-operative, mental, conditions, our inventors would naturally aspire to still higher achievements. Each year they would be able to produce many valuable inventions, which could not be used by the farm, but which could be sold by the company after being patented, for good round sums in cash! In this way it becomes evident, tnat our old members might prove the most prolific cash producers on the farm. It is even possible, and quite probable, that the sale of one invention, might bring to the company, a sum of money, more than equal to the combined pensions of the retired co-operators for one year. From this particular source, would flow an additional fund for educational work in pushing the movement before the public.

"Viewed in this light, to be retired on two-thirds pay at the age of fifty, is simply a matter of justice! When justice is done, the mission of charity is finished!

"In considering the growing interest in the insurance question among people of the outside world, we find great numbers of laboring people, and of small farmers everywhere, who are beginning to understand that it is a question of vital importance, an open gateway through which they may gain access to the broad fields of abundance. Every day, both by observation aud experience, they are taught that without the aid of some special insurance, nine out of ten who start in business fail. Also, that nine farmers out of ten, who start with a meagre capital, after twenty years of constant toil, find themselves the slaves of some money lender who holds a mortgage on the farm. These mortgages are largely the result of a hopeful struggle on the farmer's part, in a last vain effort to compete with the expensive methods of syndicate and bonanza farms.

"No wonder the average worker is anxious to discover some method of insurance, that will safeguard him against the disasters which have overwhelmed so many of his predecessors! No wonder these workers come to believe it possible, that out of a given number of say one thousand men, who start in life without capital, except such as they possess in ordinary health and strength; at least fifty per cent are liable to die in the poor-house, or in some way become helpless dependents on charity! Against such an alarming proposition, the average optimist or plutocrat, cries out, impossible! No, No! In this Republic, such things could never happen! Besides, how preposterous! Don't you know, that the general prosperity of the country was never greater than now! Why the wealth of the nation is growing at a marvelous rate! Never before, were fortunes made so easily! The way is open for every industrious man; no matter how poor he may be at the start. If people come to want in the midst of such golden opportunities, they have only themselves to blame.

"By way of an answer to these optimistic assertions, let us apply the figures collected by Prof. A. G. Warner, published in his 'American Charities.' In this book he has tabulated the results of fifteen investigations, both in this country and abroad, into the actual causes of poverty. These investigations embrace over one hundred thousand individual cases, found in the cities of Baltimore, New York, Boston, Cincinnati, London, England, and seventy-six cities in Germany. In the causes of poverty stated, eleven per cent are due to intemperance, ten and three-tenths per cent to other kinds of misconduct; while seventy-four and four-tenths per cent are due to misfortune, such as poorly-paid work, lack of work, sickness, etc. Here, we have actual proof that seventy-five thousand in the ranks of this vast army of poverty-stricken people, were reduced to such straits, by causes which they could not control.

How dreadful the significance of these terrible figures! What a blot they become, on the fair page of progress achieved by the nineteenth century! What a warning to the people of the twentieth! What an indictment against existing, social, and industrial conditions! What argument could be more convincing, or demand more imperatively, the immediate adoption of co-operative methods, which offer absolute insurance against the recurrence of such calamities?

"As relating to the insurance question, and by the way of a contrast between competitive and co-operative methods, let us consider the following statement.

"We learn from statistics, that for the family of a skilled workman of the better class—a family of five persons—the average annual cost of living is $420. This includes food, shelter, raiment, fuel, laundry, light, water, medical attendance, medicine, education and recreation.

"Under the competitive system, to earn this sum required, on the part of the adults and such of the children as were able to work, the continuous toil of three hundred days, twelve hours long—counting the possible workers of the family as three, and the labor day as twelve hours long—we have in the aggregate, say eleven thousand weary hours of this nerve depressing labor. A labor often performed in the midst of the most repulsive and unsanitary conditions; to which the toilers were constantly goaded by the cruel spur of necessity. This is a picture of the living expenses and daily working life of a family of the superior class, far above the average among the workers under the competitive system.

"To illustrate what the co-operative system can do, let us transfer the account of this family, to a co-operative agricultural colony like this. On the basis of three hundred days of labor annually, we should have daily for the two adults—the children being in school—six hours of productive labor and two hours of educative labor, an aggregate of four thousand, eight hundred hours, of work for the year. This work would be separated by such generous periods of rest and recreation, and performed amidst such pleasant surroundings, that the worker could truthfully count them as so many hours spent in necessary healthful exercise.

"As a result of this labor, we could place the annual income of the family at $1,800. All available, for providing the very best of food, shelter, clothing, heat, light, laundry, hospital service, medical attendance, medicine, education and amusement. Also superior social surroundings, with increased facilities for being well born; with educative advantages, embracing a higher order of intellectual amusements, art-culture, musical training, and industrial skill.

"In addition, the family would enjoy a savings account of generous proportions, represented by the constantly increasing value of the farm, its stock, crops, buildings, store and goods, material, machinery, industrial plants, orchards, vineyards and forests.

"Still better! They would have savings in the sinking fund, providing land, and homes for their children and grand-children in a long line of future generations.

"Best of all! This family would have savings in the insurance fund, providing for an old age of ease and comfort, free from care, sweetened and brightened by leisure, travel and the refinements of study, art and music!

"In striking a balance between these two accounts, we discover a difference in favor of the co-operative system, with its magical insurance, which is wider, deeper and more startling than the difference between the illustrations of Dante's Inferno, and the descriptions of Milton's paradise!

"A careful study of this insurance question, has taught our people many valuable lessons. They have learned to consider from a new standpoint, the object and purpose of life, and the amount of work necessary to support that life.

"They have learned that poverty is a needless crime against progress, which can and must be abolished!

"They have learned, that in these days of general prosperity, marked by a wealth of labor-saving machinery, never before dreamed possible, co-operation has demonstrated, that an average of but six hours each day, devoted to farm work, will abundantly supply the means which will yield them, the highest advantages of birth, education, amusement, and everything necessary to a healthful enjoyment of life.

"They have learned that the true purpose of work, is not to make and hoard money; but to secure these advantages for themselves and their children.

"They have learned that money is not a necessity; that it is only the means to an end. They have learned that confidence in each other, among members of a co-operative colony, working unselfishly together, largely takes the place of money.

"They have learned that practical education equips

them with a knowledge, of how to deal justly with each other, in all the social relations of life.

"They have learned that the pathway which leads to success, in winning the largest measure of all these advantages, is reached by adopting unselfish methods, which will insure the welfare of all. They have learned that this condition may be attained by building up co-operative systems that furnish renumerative self employment, and at the same time enables them to enjoy free access to the natural sources of life.

"They have learned that this free access cannot be secured, without first obtaining permanent control of the necessary tracts of land, not less than ten acres per capita. They have learned that these tracts should contain at least five thousand acres, in order to properly support an industrial co-operative colony of one thousand people.

"They have learned that the social, ethical and intellectual advantages offered to the individual, by this co operative colony life, are even greater than those relating to the question of finance.

"They have learned, that when selfish distrust of each other is once banished from the minds of the workers by the force of repeated examples of co-operative success; then, it will be practical and easy to organize the farms and farm laborers of this Republic, with its army of the poor and the unemployed of every class, into systems of co-operative farm villages, or similar industrial associations.

"In this knowledge our people rejoice! They are filled with an unselfish desire to spread the good news broadcast! Can you, my dear Fern! imagine for them, a purpose in life more noble or more worthy?"

"No, my dear Fillmore! I cannot! So eloquently have you stated the case, that the outlook for the future is glorious! How graphically you have pictured the growing importance of this question of insurance! I am amazed, and more deeply interested than ever! I never before dreamed it possible, that the co-operative farm could offer so much defense against the calamities of life, which grow out of the pinching pressure of poverty!

"The scheme for providing for the members of the Mother's Club, and for retiring co-operators at the age of fifty, meets my enthusiastic approval! I am sure it will commend itself to the workers and thinkers of the world! To me, it seems admirable, from every point of view!"

CHAPTER XXXV.

THE MOTHER'S CLUB.

"MARK it well, Fillmore! I have now reached a very important question. What have you to tell me about stirpiculture, as a part of the co-operative farm movement?"

"As a basis for the preliminary work, we have been following carefully, the suggestions of your father, Fennimore Fenwick. You will remember, my dear Fern, that they were to the effect, that the children of the farm, should be the crowning glory of all its products; that it should be the province of the corporation to provide for the children of the co-operators, every advantage of favorable pre-natal conditions, birth, unfoldment and education, that money could procure for the wealthy. Therefore, that ideal environments for mothers and motherhood, must be created and maintained.

"In order to carry out these epoch-making ideas, such of our matrons as are willing to assume the conditions, responsibilities, and cares of motherhood, are relieved from all farm work, at any time they may chose. However, much of the work is so enjoyable, and affords so much pleasant exercise, that many of them become volunteers. Meanwhile, they are paid regular wages from our insurance fund. With this abundant leisure and freedom from care, they are prepared to become zealous workers in the Mother's Club.

"Our Mother's Club at Solaris, was organized by Gertrude Gerrish, as the fulfillment of a long cherished dream. She has reason to be proud of her work! Like that other Gertrude, made so famous by Pestalozzi's charming story, Gertrude Gerrish is a born teacher, an ideal mother, one of nature's noble women. Much of the success attained by the club, is due to her wonderful power as a leader. Her enthusiasm is infectious. It has carried all obstacles before it. To this self appointed task, she has given her best energies, a rich harvest of ripe experience, with its fruitage of earnest thought, radiant and glowing with the genial influence of her sunny temperament, and withal, rendered more potent, by an overflowing love from the deep fountain of her great mother heart. Is it a matter of wonder, that she is such a general favorite with club members! Her word they accept as law. Her suggestions as commands.

'To Gertrude Gerrish, motherhood was a holy and sacred office, which demanded from its devotees, a season of careful preparation, and a thorough knowledge of the physiological and psychological laws, which govern that life-evolving function, that crowning glory of womanhood. She seemed to be inspired with the idea, that progress has ordained, that unwilling, ignorant and accidental mothers, must be replaced by those who are predetermined, properly educated and fully prepared. These ideas, she has endeavored to impress most forcibly, upon the minds of all club members. She has also taught them the importance of maintaining joyous, healthful, mental conditions; consequently, of carefully avoiding all emotions of selfishness, cruelty, anger, envy, or melancholy. In this connection, for the purpose of creating in the minds of our club mothers, as many good and pleasurable emotions as possible, and of repeating these anabolic emotions so often, that they may become dominant during the entire gestative period; Gertrude Gerrish has wisely planned for them, a great deal of open air exercise, study and amusement.

"The study of botany, and botanizing parties, have become very popular. These prospective mothers, have quickly learned how to amuse themselves, by combining study with pleasure. When organized into congenial outing parties, almost every fine day they may be found, seated in the luxuriously appointed motor carriages which belong to the club, ready for a lively spin away to the woods. This gives them an opportunity to enjoy the pure air and bright sunshine, the wide, undulating landscape, tinted by the exquisite coloring of every flowering plant, shrub and tree. How delightful to them, is the restful green of dewy meadows; the sweet music of birds, the charming chatter and playful antics, of the swift-footed squirrels! How grateful, the leafy coolness and bracing ozone of the forest; the dancing shadows of its deep glens, with their garnered treasures of mosses and ferns! How inspiring, the merry tinkle of the clear streamlet, swiftly flowing over its rocky bed; or the louder roar of the rushing waterfall, where drooping boughs glisten and sparkle with spray-laden foliage! All these, are nature's matchless charms, which appeal to our young mothers in their best moments, their most responsive moods; banishing all thoughts of evil, awakening in their hearts, new spiritual impulses, feelings of worshipful adoration; emotions of the highest and purest order. Than this, nothing could prove more helpful in maintaining perfect conditions of mental and spiritual serenity.

"Inhaling the pure, invigorating air of the country, far from the dust and filth, the smoke and poisonous gases, the turmoil and strife, the ceaseless din, the selfishness and sin of the great city, close to the fostering bosom of mother earth, under a broad dome of blue sky, bathed in floods of golden sunlight, exulting in the exuberance of perfect health, these grateful young mothers, realize how much they owe to the co-operative farm movement, for surrounding them with such ideal conditions of life.

"They realize, the great, good fortune of children, who are born and reared in the midst of such delightful environments. They perceive, with a keen sense of sorrow, that children who are born and bred away from these rural conditions, are robbed of more than one-half their natural rights. They realize, more than ever before, the filth, the misery, the squalor, the fetid air, and the unsanitary conditions, of our great cities. They shudder, when they contemplate, the bitterness of the misfortune, the cruelty of the deprivation, of the great mass of children, who must be born and bred in the midst of such depressing, unhealthy surroundings. They know intuitively, that only a puny, sickly, half-developed race of people, can come from such a sad birth. Under such circumstances, they do not wonder, that fully one-third of the human family, die in infancy.

"Indoors, the handsomely furnished, beautifully decorated club rooms, which are located in the kindergarten building, offer the maximum of elegance and comfort to club members. There, in harmonious groups, they may engage in conversation, study, writing, musical exercises, and other varieties of club work. The esthetic tastes of the members are quickened, and their pleasures much enhanced, by the fine display of oil paintings, water colors, pencil sketches, etchings, and photographs, which have been hung on the walls, by admiring friends from the art and photography clubs. It has been the chosen work of the last named club, to supply the center tables in the reading rooms, with a series of large portfolios, containing a choice collection of finely finished, beautifully mounted photographs. This collection is varied, unique and valuable; and withal, exceedingly interesting. It embraces artistic copies of the world's finest statuary, pictures of eminent men, noted, historic buildings, rare landscapes and most picturesque scenery. These, supplemented by an abundant supply of choice books, furnish excellent conditions, and a most fascinating incentive, for a harmonious, satisfying, self-culture, of the highest type. Under the able leadership of Gertrude Gerrish, the interest shown, the enthusiasm awakened, and the progress achieved, is something remarkable.

"Thus prepared, the members find themselves on a higher mental and spiritual plane of existence, where they can appreciate the possibilities, of what may be accomplished by true motherhood, as a regenerator of society. They can understand the significance of the great lesson taught by history, which is, that all progress for the race, depends upon the elevation, education and refinement, achieved by woman. With quickened vision, they can perceive, that with the dawn of the twentieth century, comes the beginning of a new cycle in the life of the planet; the commencement of woman's golden era! In the higher light of such a vision, they become aware, that they must strive continually, for more wisdom, that they may reach a higher consciousness of individual responsibility, as keepers and guardians of the sacred temple of human life.

"In the preparatory work for a progressive parentage, club members are taught, that prospective fathers and mothers, must become familiar with the sciences, the industrial, and the higher arts, if they wish their children to inherit, whatever intellectual progress, they as parents, may achieve. The new psychology, with a better knowledge of nature's evolutionary methods, declares, that these trained intellectual attributes, may be transmitted to offspring, if the parents are willing to prepare themselves, to respond to the demands of natural law.

"In the domain of more practical club work, the members are taught how to prepare the diet and clothing, which may be necessary for the proper care of healthy nursing mothers and infants. They are also taught the hygiene and physiology of motherhood; in addition, as much as possible, about the laws that govern the procreative body of woman, when it becomes the temple of evolving life. In connection therewith, they are instructed to observe closely, the initial and pre-natal conditions, which dominate this primal stage of embryo life.

"As a result of this comprehensive course of training, our young mothers soon find themselves, inspired by a hypnotic wave of enthusiasm, which is sure to follow many days of pleasant association, discussion, and systematic study. Stimulated by this enthusiasm, and aided by the potency of co-operative thinking, they endeavor to discover new avenues, through which they may reach and maintain, better physical, mental and spiritual conditions, which shall bring them into a more perfect harmony, with the laws of unfoldment which govern planetary evolution. The success, which has rewarded their efforts in this direction, has far exceeded, even the ambitious hopes of Gertrude Gerrish.

"For the purpose of preserving a series of valuable records, for the benefit of this and coming generations; club members are urged to put in writing, such ideas as may come to them, as the result of individual thought, or from co-operative study, discussion and observation. These papers are carefully condensed, sifted, classified, and placed in proper record form, by the editing committee of the club. This committee, is also instructed to prepare short extracts, essays and descriptive articles relating to club work, for publication in the mothers' column of the Solaris Sentinel.

"This outline sketch, my dear Fern, will give you some idea of the scope of the work, in which, I know you are greatly interested. In brief, it means a practical illustration, of the use of scientific methods, for improving the race. The club hopes to give a satisfactory answer to the great question, of how to be well born. It will strive to convince the world, that the time has arrived, in which the twentieth century demands the immediate introducton of a scientific system, for the thorough breeding of children as a fine art. The art of all arts! The highest of all possible achievements!

"Hitherto, the world's people, in trying to accumulate riches, or to escape the poorhouse, have had neither time nor inclination, to consider this most important of all questions. As a matter of fact, greed for gold has become so dominant, human life, so cheap, and its progress through culture, held in such low estimation; that it is not unusual, not even a matter of comment, to hear of a wealthy stock-breeder, who willingly pays from ten to twenty thousand dollars a year to the trainer of his horses; while he grudgingly pays five hundred dollars a year to the teacher of his children. This would indicate, that the demand for a change is imperative. The great wave of evolutionary progress, is fast rising to a flood tide! The selfish, commercial spirit, born of the competitive system, must soon give way for something better! The advent of a system of unselfish, co-operative farming, which proposes to unite a rational agriculture, with a scientific stirpiculture, offers opportunities for substantial progress, and a new hope for the coming race."

"This is exceedingly interesting, Fillmore! What additional work, has Gertrude Gerrish planned for the club members?"

"A great deal more than I have time to enumerate, just now! However, by the way of an illustration of her ingenius methods, and also, of the great variety of the topics introduced, all of which really belong to the work, as an integral part of the movement. I may mention the latest scheme introduced by Gertrude Gerrish, which proposes to increase the average length of human life, by giving to children as a birthright, well developed vital, physical, and mental organizations. This, she claims, is the only true ground work, for real progress in the right direction. The scheme has proved a popular one. It has so aroused the zeal and enthusiasm of the club members, that they write, think and talk on the subject, with an inspiration and eloquence quite surprising. As a result of the remarkable interest awakened, they have diligently read books on evolution, physiology, psychology, vital statistics, physical culture, and a great number, on the general subject of health. In this respect, the work of the club as a promoter of longevity, may well serve as an object lesson, for the hundred-year clubs, that have been organized during the past ten years, for the purpose of checking the alarming increase of suicide clubs.

"Touching the question of suicide, as an enemy to longevity: In discussing the subject, many members of the club maintain, that it is an imperative duty for them to give the world a new cure for suicide. They would offer its would-be victims, such a tempting array of the meanings, purposes and opportunities, for gaining wisdom, which may crown every rightly conducted, harmoniously environed life; making it so busy, so absorbing, and so happy; that there would be no room, for the morbid hallucination of a suicidal desire. This proposition is based on the presumption, that all suicides are possessed with an insanely erroneous idea, regarding the true object and purpose of human life. After the passing of a few generations, under the wide-spread reign of co-operative stirpiculture, with its hosts of mothers' clubs, suicide will soon become an utter impossibility.

"In the ever broadening scope, of progressive kindergarten training, our young mothers have wrought their most important work. A work, which reflects on the club, a great deal of well-earned credit. As centers of the first and second-year nursery groups, in their cargosita excursions around the great hall, for the purpose of sight, color and image training; the service rendered by these mothers, has proved invaluable. As teachers, assistants, directors and leaders, in the third and fourth-year groups, while engaged in exercises and games, which have been devised and instituted, for the purpose of sense training, science training, and science recreation; in addition to the ordinary kindergarten course; their excellent work, has justly excited the pride of the colony.

"In conclusion, my dear Fern! I must tell you something about 'The club babies,' as they are proudly designated by the members. They are very bright and beautiful! In fact, they seem born with a consciousness, that it is their peculiar privilege, to commence the study of life as a fine art, at its very threshold. They are the zealously guarded treasures of the club, and the pride of the farm! They give a glorious promise, that they will prove worthy leaders, of a coming host of dominant thinkers, which are to be given to the world, by the mothers' clubs of the next quarter of a century.

"As champions and exponents of the true object and purpose of human life. these thinkers will be armed with a wonderful potency, with which to overcome and conquer, the selfish reign of the competitive system. A cruel system, which has proved the very incarnation, of 'Man's inhumanity to man,' causing countless millions to mourn! In this great work, they will be inspired, by the high purpose of replacing its evil, poverty-breeding dominancy, by an unselfish, co-operative system, a union of spiritualizing, educative, stirpiculture and agriculture, which shall insure a higher civilization, and the perpetual reign of peace and plenty for all mankind."

"What you have told to me so charmingly, Fillmore, is almost too good to be true! How eloquently, and how interestingly, you have described, the scope and work of this wonderful club, with its gifted leader! I hail the advent of this club, as one of the most important results, achieved by the Solaris Farm Company! I am delighted, with its thorough organization, broad plans, high aims, earnest work, and the remarkable enthusiasm, of its members! They represent a cause, which is dear to my heart!

"The question, of how to be well born, is to my mind, the foremost question of the day! A question, which demands universal consideration! This twentieth century union, of agriculture and stirpiculture, this scientific, systematic, generation of the race as a fine art; which has been so well demonstrated, by the surprising work of these enthusiastic young mothers, is something to be proud of! The good, which must follow the work of this club, cannot now be estimated. The one hope, for the regeneration and final salvation of society, is centered in the mothers of the Republic! Nothing, is so well calculated to impress the importance of this grand truth, on the minds of the people, as the practical work of an ever increasing host of mothers' clubs.

"In their devotion to the Republic, these mothers are patriots of the purest type! They have arisen to such spiritual heights, that they may fearlessly proclaim the láw of motherhood, for the sons and daughters of the new Republic! They have demonstrated that this law declares, that a worthy mother of the new Republic, must be absolutely free! She must be free, religiously, mentally, socially, physically, and financially! Thus unshackeled, she may be properly prepared, to bear a race of children who are endowed by birth, with the incarnate spirit and genius of true liberty. Such liberty, as shall become the talisman and watchword, of the model Republic of the twentieth century. A Republic of peers, of intellectual giants! The very flower of spiritual unfoldment! The highest order of civilization! Under the starry flag of such a government, neither slave, nor pauper, nor criminal, shall be found to cloud with shame, the fair escutcheon of true liberty!

"I shall endeavor, before leaving Solaris, to meet with the members, by attending some session of the club. I shall then take pleasure in restating these ideas, as an expression of my appreciation of the great work for humanity, which they have so successfully inaugurated.

"To Gertrude Gerrish, that noble woman, with such a magnificent talent, and so loyal a heart; who has won my deepest gratitude, my undying respect; I must pay the tribute of my admiration, by taking her lovingly to my heart, as a sister woman, whose wonderful ability, as a thinker, organizer, and leader, has made me proud of my sex."

CHAPTER XXXVI.

THE CO-OPERATIVE FARM AS A FACTOR IN THE CAPITAL AND LABOR PROBLEM.

"I AM curious to know, to what extent co-operative farming will effect the capital and labor problem. What think you, Fillmore?"

"No doubt the effect will be very marked. Many of the solutions arrived at in experimenting with the insurance question, will apply with equal force towards a final solution of the capital and labor problem. The toiler once having been taught the art of self-employment, that will furnish him superior conditions for a perfected healthful enjoyment of life, with all of the advantages for himself and his children that money can buy for the wealthy; can never again become the working slave of capital. He has learned, by a practical lesson, very similar to the famous 'Gurnsey Market House' exploit, that labor unaided by capital, can produce an abundance of things which go to make up the wealth of the nation, the community or the individual; while capital unaided by labor can produce nothing.

"In searching for a remote cause for this ever growing warfare between capital and labor, which has so long vexed our Republic; and which, even now, threatens its final disintegration; we soon discover our arch enemy, the competitive system, as the party responsible for the mischief. This fact becomes more apparent, as we consider, that from the beginning of the historical period, people in a fierce struggle for existence, have been compelled by the competitive system, to wage a brutal, relentless warfare with each other. Always the stronger, against the weaker. In this wicked war, millions of human lives have been sacrificed to the fiery moloch of selfish greed.

"The older the civilization the more fiercely has the war been waged; until to-day, thousands among the lower classes everywhere, dwarfed and embittered by a hopeless struggle to sustain life, in a ceaseless combat with competing foes on every hand; spurred to a frenzy of fury, curse the day which gave them birth. Why should they live only to suffer? With moral natures starved and withered, they declare that all justice is a mockery, all honesty, a myth! They have lost faith in God, and confidence in man! They care not for the needs of posterity, or for the nemesis of a future existence! In this desperate condition, they either commit suicide, or become an easy prey to the temptation, to join the outlaws in taking the world by the throat. From such material is formed the dregs of society, that lower social strata of living dynamite, that constant menace, which threatens in the near future, to destroy all civilization which rests upon it. This is a typical piece of the handiwork of the competitive system, a system in which the roots of society today are grounded.

"Once seriously considered in this light, how can any sane person, who believes in an All-Wise Creator, in justice and mercy, in a common brotherhood for humanity, ever again defend the wickedness, of a society based on the selfish cruelty of such a system? What treatment may unorganized, unprotected labor, expect from this system?

"Hitherto, fortunately for the progress of the world, the laborers of this Republic, have enjoyed more of the advantages of life, than those of any other country. With better wages and shorter hours for work, they have been able to educate themselves and their children, to a degree that would fit them to become good citizens of the Republic. A republic which for its continued existence, depends on the integrity, ability and intelligence of its working units. As such, our laborers have proved themselves the best in the world. Now, alas! The whole industrial situation is changed by the swift dominancy of the competitive system, with its ever increasing brood of trusts, which have swallowed up all natural opportunities, and monopolized all the leading business enterprises, of this hitherto progressive nation.

"The people of the Republic are divided into two classes; the employers, and the employed. The invention and introduction of new and expensive machinery each year, augments the power of the trusts, to control the markets and the industrial situation. By the same means and at the same time, they are fast reducing the number of employers, and increasing the number of those who must seek employment. Under such circumstances, each year the fate of the worker in any class, either skilled or unskilled, grows more desperate. He becomes more completely the slave of the trusts or capitalists who own the tools and who monopolize the industries. The larger the dependent family of the worker, the more abject the slavery, and the less his power to resist a constant reduction of wages.

"In the efforts made by organized labor unions, to resist this tendency to reduce wages, we have both the cause and the beginning of the war between capital and labor With a courage and patriotism worthy of the days of 'Seventy-Six,' this war has been waged by the toilers, with a determination to maintain rights guaranteed to them by the constitution of the Republic. A right to life, liberty, and the pursuit of happiness. A right to labor and to enjoy the fruits of their labor, by having free access to a reasonable share of the natural advantages belonging to the public domain.

"In this heroic struggle, so sturdily maintained during the past twenty-five years against the competitive system and its well trained hosts ; the campaign, which has been marked by many mistakes, followed by frequent defeat and disastrous failure, has always proved successful as an educator, both for the toilers and the great middle classes, who sympathized with them. On the other hand, alarmed by sudden success, achieved by the disruption of long-lived business methods, and the loss of confidence in exchange values, on the part of the public in consequence of this disruption ; the generals of the competitive system, aided with but few exceptions, by the press, university and pulpit, have shrewdly endeavored to evade responsibility, for the disastrous panics which have followed such revolutionary methods. These panics have left the country disturbed and embarrassed, by armies of unemployed men.

"In the same line of tactics, these competitive leaders, have endeavored to confuse the question, and to mystify the people, by raising the cry of overproduction ! The inexorable law of supply and demand ! The impossibility of our manufacturers longer competing in the markets of the world, against the cheap products of the pauper labor of Europe, while they are obliged by the unions, to pay such exorbitant wages here. This cry has grown more insistent, with each succeeding year. Nevertheless, the fact still remains, that but for the continuous opposition of the united labor organizations, long before this time, the wages paid in Europe, would govern the price of labor in this Republic. What then would have happened to our workers, the basic units of our government? Fortunately, the campaign of education still continues! The people at large are just beginning to wake up to the importance of the labor question! They have studied it carefully and earnestly. They have learned that in productive labor, muscular effort is a mental demonstration.

"They have learned, that the products of the skillfully educated, intelligent, refined, moral, self-respecting worker of this Republic, can successfully, compete with the inferior products, of a less intelligent or pauperized labor of any country, in any of the markets of the world. No matter how high the wages of the former, or how low the wages of the latter may be.

"They have learned, that the demand, in any market for a superior article, will always drive out the inferior.

"They have learned, that the question of the unemployed, is a question of the utmost importance, which demands the immediate attention of all patriots. They have learned, that the unemployed we shall have with us in ever increasing numbers, so long as the competitive system shall last.

"They have learned, that not one from the ranks of the unemployed, can again become a worker, without paying a handsome bonus for the privilege, by allowing some one to pocket the lion's share of the profits he may be able to earn

"They have learned, that when society encourages conditions, which cause the laborer to look upon any calamity as a blessing in disguise, because it offers work for the unemployed; that society, must be reorganized.

"They have learned, that whenever an industrial system produces conditions, which make the laborer see only disaster for his individual interests, in every labor-saving invention which may be introduced; such a system, must be superseded by a better one.

"They have learned, that the competitive system, by the very nature and terms of its organization, obliges its followers to be selfish, cruel, heartless, unmanly and unpatriotic. They have learned, that its reign has become so dominant, that it justifies a recent writer of most excellent wit, who declares that 'Man by birth, education and training, has become so essentially selfish, that no preaching has any effect upon him, if it does not advise him to lay up treasures for himself somewhere.'

"They have learned, that the dangers which most seriously threaten the perpetuity of our Republic, do not come from the clamor of dissatisfied laborers, who are wrongfully accused of law-breaking; but, that these dangers do come, from the lawlessness of capital, and the anarchy of corporations.

"They have learned that so far as the interests of the working units of the Republic are concerned, or care for its continued existence as a representative government; the press, the university, and the pulpit, have all been syndicated and censored by the competitive system to such an extent, that they can no longer be trusted to furnish teachers, leaders, and guides.

"They have learned, that the only safe course is, for the people to depend upon themselves, to develop and establish a new social and industrial order, from which shall spring a class of incorruptible leaders and statesmen, whose pure, unselfish motives, dominant, evenly developed minds, and superior ability, shall mark them as fitting rulers for a more perfect Republic. Such a Republic as shall meet the demands of a twentieth century progress.

"They have learned, that the remedy indicated is a change to an industrial system, that will secure to the laborer an equitable share of the benefits, which follow the introduction of labor-saving machinery. Under such conditions, the laborer himself, having more leisure and unexpended vitality, will be stimulated to increase his available resources by cultivating his brain capacity for invention, thereby largely increasing his power to produce.

"After many years, the rank and file of the workers in the labor unions, have learned, that self-employment is the key to the situation. Although late, they have learned, that if all the money wasted in unsuccessful strikes, had been invested in the purchase of choice locations, undeveloped mines and mineral lands, and in the erection of manufacturing plants, the labor question would now be a thing of the past. They would be masters of the situation, to whom the capitalists would be glad to offer such a liberal system of profit-sharing, as would practically make the workmen self-employed, by reason of a part ownership in the enterprise they labored to exploit.

"Finally, and most important of all; they have learned that all manufacturing industries, naturally grow out of agriculture. That the success of one, is the measure, for the success of the other. That they must co-operate to such an extent, that a constant, healthy growth of both, may be maintained.

"They have become convinced of the imperative necessity for this equable, co-operative, progress, by a careful study of the threatening conditions which obtain, in countries where agriculture has declined; and where manufacturing industries have become abnormally predominant. In such countries, the food supply at once becomes a question of daily, nay of hourly importance. It must be imported from distant lands, subject to the tax of insurance, import and export duties, freight charges, and commissions. Under such adverse conditions, available supplies for but a few days only, stand between the toiler and gaunt hunger. Any catastrophe which may happen to already congested lines of transportation, will percipitate a famine. Then prices would go up with a bound. The constant menace of such a possibility, always serves to keep food-prices above the natural level of a fair profit. On the other hand, in countries where progress in agriculture and manufacture goes hand in hand; a constantly increasing home market for manufactured products is steadily maintained. A most important consideration! At the same time, the industrial centers have the advantage of the immediate vicinity of abundant food supplies, which are not subject to the vicissitudes of traffic or transportation, or to the tax of much handling.

"In considering these things, the minds of a great majority of the laboring people, have been prepared to accept, the conclusion, that the great question of the hour is, how to open the way for every worthy worker to become his own employer. The co-operative farm opens the way. Therefore, it is to these self-educated toilers in the ranks of the labor organizations, that the manifest advantages of co-operative farming will appeal most successfully. If properly approached, a majority of them would be, not only willing but anxious for an opportunity to give this new system of co-operative agriculture a thorough trial.

"Having once become practically interested, these people would soon learn to consider the object and purpose of life from a new standpoint. From this new concept of the meaning and necessities of life, they would perceive that it did not require the hoarding of much wealth, in order to satisfy them. The insurance system in providing for the wants of old age, would forever banish the haunting specter of a pauper's death in the poor-house. They would then realize that money, was not so precious as a human life! They would clearly understand that money was an absolute necessity, only to those under the competitive system who had lost confidence in each other, and faith in the fact of a common brotherhood for humanity!

"They would soon respond to happier surroundings, in every way so conducive to a natural, soul growth, and to the harmonious unfoldment of the individual from within. In this unfoldment, a new meaning for immortality would come to them. Spiritual law would become operative. It would teach them that, as immortal beings, as cosmic units of the larger cosmos—The Great Over Soul—they could not become totally depraved, even under pressure of evil conditions of the most degrading character; no matter how much their spiritual natures had been stained or starved.

"With this new standard as a guide, there would come an inspiration to strive for the attainment of a higher, purer, better life. A life more in harmony with the design of an All-Wise Creator! Angry, antagonistic feelings, against hitherto competitors, would disappear. The world would wear a smile instead of a frown! Brotherly love between man and man, would become the rule in place of the exception! Gold would lose its charm! Avarice would pass away! Selfish instincts, born of bitter years under a cruel system would soon follow! Long dormant, spiritual natures would be awakened! A new spiritual growth would take place! A vastly wider, mental, and spiritual horizon, would be added to the wisdom of the individual! In the light of this wisdom would come the discovery, that the virtue of right living, bears the seeds of a perpetuity, which begets true and lasting happiness! An overwhelming answer in the affirmative, from every point of view, to the question, does it pay to be unselfish?

"With higher ideals of life and its duties, these physically, mentally, and spiritually emancipated toilers, would find themselves prepared to co-operate most effectually, in establishing and maintaining any social and industrial evolution, which the best interests of the people and the Republic might demand.

"From this presentation, my dear Fern! you may imagine how important and desirable it is, that these two powerful industrial forces should become harmoniously united in working for the interests of a natural progressive evolution. Against such an invincible combination, the hosts of the competitive system might not hope to prevail! Once thus united, each co-operative farm would then become the nucleus of a new industrial organization, capable of such unlimited expansion and perfection as the needs of surrounding communities might be able to sustain.

"As this twin series of giant industries continued to grow and expand, the ways by which they might co-operate with mutual benefit, would continue to multiply. In political matters such a combination would prove remarkably strong; first in the township and county; later, in state and national legislatures, where it would soon be able to demand and push forward favorable legislation, and also to strangle much that might threaten to prove adverse. In such efforts, would come opportunities for introducing to the arena of public life, an abler, nobler, purer class of young men; who, born of a better social, industrial system, by reason of superior conditions for birth and training, would be properly endowed with that inspiring patriotism, sterling integrity, and commanding ability, so necessary to maintain the dominancy and perpetuity of the Republic, as a government of the people, for the people and by the people."

"Bravo! Well done Fillmore! Your statement of the subject is grand, indeed! The eloquent summing up, forms a fitting climax in answer to my last question, the closing one of the series. But, as much as I admire and appreciate its general excellence, you must allow me to suggest one criticism. Do you not think Fillmore, that you put the case rather too strongly, when you place the press, the university and the pulpit, so completely under the control of trusts, or the leaders of the competitive system? Would they dare to do such a thing?"

"Bless you my dear girl! They are capable of doing anything! So far as the trusts and the competitive system are concerned, I have stated the case very mildly. Not one-half of the story has been told. Let us probe this question a little deeper.

"What is a trust? It is the highest form of monoply. It is a nest of corporations, laid and hatched by the competitive system! It has neither conscience to hold it in check, nor soul to be damned! It dares to do anything! Indeed! It is formed for the sole purpose of making money. Nothing is allowed to stand in the way. Born of the consolidating pressure, which marks the competitive system, it seeks to monopolize all of the advantages of that cruel system, without incurring its penalties. Once thoroughly organized, and armed with the almost unlimited power of its enormous capital; the trust immediately commences the wholesale destruction of all opposing industries or interests. In pushing this work, it regards neither the equities of commercial law, nor the vested rights of others. Securely protected by its monopoly, this modern juggernaut in the commercial world, rolls remorsely onward toward its goal of wealth. It cares not for the safety of worshippers, friends or foes. If by chance they represent competing interests, they must either leave the field or be crushed. There is no alternative! There is no escape!

"A few of the leading trusts, those most completely representing the competitive system, have recently become so defiant, so audaciously bold, that they are prepared to undertake, to consolidate the business of the whole earth. They will stick at nothing! They have the gorge to swallow one government or ten! It matters little to them! Like the ring of conspirators, in Donnelley's 'Ceaser's Column,' a few of the leading spirits, of these daring trusts, are secretly plotting in Gotham! Just at present, they have their eyes fixed on the all-powerful money question. The vision seems a pleasing one!

"What is that question, which so completely absorbs the attention of these people? Can it be possible, that the mills of the competitive system will grind up rich bankers, as unconcernedly as they do the helpless poor! They surely will! The plot grows and thickens! Let us give it close attention. Let us watch these people. Keeping in mind meanwhile, that hitherto, the bankers of the country, have complacently considered themselves masters and kings of the financial situation, whose thrones were secure for all time. Strongly intrenched behind well-filled money bags, they have felt themselves safe in helping the trusts to fleece the public. Now they are becoming alarmed. They are shaking in their fifteen-dollar boots! They behold that dreadful handwriting on the wall! In giant letters, seemingly towering forty feet tall, these bankers read the doom, which the trust conspirators are now preparing for them. They catch the frightful significance of the question, which the trust leaders are discussing. It is this. Why should the business of the United States, support such an army of banks? More than ten thousand. We know very well, that the entire money transactions of this country, could be handled more safely, more swiftly, and more cheaply, by one grand central institution. With one voice the conspirators exclaim! Let us form a pool! Let us consolidate the whole business, into one magnificent money trust! Let us select, say twenty-five, of the brainiest bankers in the business! Let us give them fat salaries, and make them superintendents of the financial agencies, now called banks. Counting the whole number of banks, both public and private, as ten thousand, with three professional bankers to each one, the result would be a total of thirty thousand bankers. Of this number, we could reduce twenty-nine thousand, nine hundred and seventy-five, to the station of bank clerks. Let us pause for a moment to contemplate the result! What enormous savings would accrue, by the introduction of such a wholesale scheme of consolidation! These savings would be ours! Intoxicated with the brilliancy and the hugeness of the idea; the conspirators with one impulse, spring to their feet, with outstretched hands they form a ring, they execute a round dance extraordinary. While thus engaged, they gaily shout, 'There is millions in it for us!'

"No wonder the bankers are alarmed! With the exercise of one-half of their usual cunning and foresight, they should have scented the danger sooner. No doubt, they were so engrossed by the fascinating game of money grabbing, that they were wholly blind to danger, as the result of the combined audacity and perfidy of their former partners. They have evidently failed to learn one plain lesson, which is taught by the logic of events. It is this. When once fairly started, the process of the larger corporation, swallowing the lesser, goes forward with such an ever-increasing rate of speed, that it soon overtakes and gobbles up banks and bankers.

"At this point, it is pertinent to propound the following questions: If this is a Republic? If the people are the government, and the government is the people? And if the consolidating business, is so good and so profitable for the trusts? Why, should not the government, own and run this giant central bank? Why, should it not own and operate the railroads, the canals, the shipping, the mines, the forests, and all other industries? This would give the people a chance to share equally, in the enjoyment of these enormous profits. Why not?

"What say you my dear Fern! Would it not be infinitely better, than to allow the government to be swallowed by one monster trust?"

"Better Fillmore! Far better! I am convinced! I withdraw my criticism. You have maintained your point so vigorously, that I have not the courage, to offer one single word in reply. I am ready and willing, to consider the discussion as finally closed."

CHAPTER XXXVII.

THE CO-OPERATIVE FARM TRIUMPHANT.

THE BEGINNING of the second decade of the twentieth century, saw the final triumph of the co-operative farm at Solaris. The five years of trial and probation, have swiftly passed into history. The labors of the colony, have been crowned with a rich harvest of success. A great work for humanity, has been accomplished. A grand lesson in the economics of unselfish co-operation, has been demonstrated. A kaleidoscope of new charms, of fresh beauty, of an infinite variety of change, of unexpected opportunities, of a host of new expressions, in the possibilities of social and industrial life; the culmination of untried methods, new hopes and new aspirations; have marked this victorious climax. All have contributed, to the happiness of the contented villagers at Solaris ; filling their hearts with brighter hopes for the future.

A new era in agriculture has dawned. With it has come, a new order of life for farm people. The links of social life, have become more firmly knit. New chains of enthusiastic interest, in the humanitarian work represented by the farm, have been forged by the binding associations of passing years. Ethical, industrial and spiritual life, has been unfolded, in harmony with the law of progressive planetary evolution.

As an illustration of the perfected possibilities of rural life, this suggestive and pleasing picture is well nigh complete. Verily ! Virtue has been richly rewarded, by the pure pleasure of right living ! To the truths of these things, the lives of the unselfish co-operators at Solaris, bear most abundant and convincing testimony. Happiness and contentment, reign supreme ! Social solutions, offer new fields of pleasure to a generous, progressive people, who are daily becoming better educated, more dominant as thinkers, more unselfish in all things, therefore, more virtuous.

In passing from the experimental, to a more perfect stage of co-operative life, a marvelous change for the better is noted. New factories have been built, new industries instituted, and organized. The busy hum of industrial prosperity, everywhere claims attention. Meanwhile, the demands for a better esthetic culture, have not been neglected. The interiors of both factory and workshop, have been made additionally attractive, by a more artistic, educative class of decorations. All industrial buildings, are surrounded by well-kept lawns.

Many handsome cottages, showing a great variety

of beautiful designs, cosey, vine-clad and picturesque, environed by gardens and lawns, have been added to the architectural display of the village. Order, symmetry and cleanliness, have become the established law of the farm.

Barns, stables, stock yards, pig pens and poultry yards, have been placed at a safe distance from the village. In the erection of these necessary buildings, care has been taken, to provide for the removal and sanitary dry storage, of the daily accumulation of valuable manures. Especially designed machinery, accomplishes this otherwise unpleasant task, quickly and easily. By this convenient arrangement, with a very little labor, these buildings, and the stock housed in them, can at all times, be kept healthy and clean. A most important consideration!

Everywhere, appear evidences, of the farms increasing wealth in live stock. Great herds of fine cattle, are fattening in the fields, pastures and barns. Prize collections of choice sheep, are roaming over grassy slopes. Fine droves of well grown, healthy swine, in assorted lots, are contentedly feeding in small fields of fresh clover. The large drove of beautiful, highly bred horses, is a very valuable one. The poultry yards, are filled with many varieties of fine fowls. All show the effects of careful attention, from the hands of care takers, who are both kind and skillful.

On the opposite side of the village, near the nursery, the numerous fish ponds are located. Flower bordered, island studded, and tree margined, with surfaces dotted here and there, by tiny fleets of graceful, shell-like pleasure boats. They add much to the rare beauty of this pastoral picture. Beneath the rippling surface of the clear water, in these miniature lakes, flash the shining scales of a swarming host, of the most delicious of food fishes.

Fragrant, purple and gold, the heavily laden vineyards, are growing and glowing in the bright sunlight. They give promise of an early generous fruitage. Thrifty orchards of healthy well-grown fruit trees, including many varieties, are fast coming to maturity. Waving fields of golden grain, ripple in the simmering heat of a noon-day sun, or rustle and billow with each passing breeze, under the pale light of a harvest moon. Beautiful fields of cotton and corn, are an inspiration to behold. Fine fields of vegetables, nurseries, gardens and shrubberies, with a wealth of lovely flower plots, all add to the charm of the general effect.

The extension of the co-operative system, to embrace the second farm, has been well started. Fenwick Farm, is the name chosen for this farm number two, of the series. Two years of intelligent, well-directed work, by its wide awake, industrious people, have shown surprising results ! They are constantly inspired to do better work by the hope of being able to reach a degree of success, equal to that achieved by Solaris. In this respect, the spirit of healthy rivalry, which has arisen, gives them an advantage, which the parent colony did not have. The success already attained by Fenwick Farm, has attracted widespread attention, in the surrounding communities. The effect for the good of the county, and of its people, socially, politically and financially, has been quite remarkable. The tax payers of the county, are delighted ! They have been completely won over, to the side of co-operative farming, by the force of this second example.

One of the greatest gains, which has arisen from co-operative effort for mutual benefit, between the two colonies, has been practically illustrated, in the great work of road building. These two co-operative farm villages, are now connected by a broad, smooth, well graded road. This road, ten miles in length, is margined by a wide strip of beautifully kept parking. Five miles of this parking, on either side of this magnificent boulevard, become the especial care, of each village. No city in the union, could display better taste, or greater pride, in keeping these beautiful parks, in the most perfect condition.

In order to keep the park lawns, foliage and flowers, always looking clean and bright, it becomes necessary to keep this road free from dust. For this purpose, the entire road surface, is given a frequent sprinkling with petroleum. After each sprinkling, the enormous pressure of an hundred-ton roller, soon converts the layer of moistened dust, into a hard, smooth mass of oily rock. This process is repeated until a thick, heavy, durable surface of water-proof rock, is secured. This makes an ideal road! The hard, well pounded, gravelly soil, below, gives a permanent foundation, because it is so well protected against moisture, by this broad, indestructible roof of oily rock. The wide, slightly rounded surface of the road, sheds water like a duck's back. Consequently, it is always free from mud and dust. The broad rubber tires of a great variety of freight motors, pleasure mobiles and motor cycles, do not wear its perfect surface. The very acme of pleasure is reached, in riding over such a delightful road!

After work hours have passed, the pleasure seekers from both villages, in merry congenial parties are awheel, enjoying to the utmost, the pure, sweet, flower-perfumed air, together with the soothing, restful beauty of a park lined drive, of such extent and variety, as a multi-millionaire, might not be able to command. Could anything more delightful be imagined! Is it any wonder, that people from adjoining counties, thirty miles away, come in droves, to enjoy a ride over this now famous road! In the hearts of all comers, is stirred the imitative spirit of rivalry. They return to their homes, determined to co-operate with their neighbors, at least to an extent that will enable them to build such roads for themselves. They are convinced, that the excellence of its roads, in any community, is the only sure test, which will indicate the exact degree of civilization, attained by its people.

At the village of Solaris, the universal use of Solaris brick, of the various patterns and sizes, has proved an important factor in the construction of sidewalks, store houses, industrial buildings, cottages, the hotel, the schools and the theatre. The visitor is at once impressed by the wholesome, attractive, substantial appearance, given to the town by the use of this excellent and durable brick. In this respect, the square mosaic bricks, of unique design, used in laying the broad sidewalks, twenty feet in width, which border Railroad Avenue, the street leading straight from the public square, to the railroad station, create an effect so marked that it never fails to attract attention and admiration. The symmetrical trees and well-kept parking which line this avenue, serve to enhance the pleasing effect.

The artistic skill acquired by the people of Solaris, in the making and laying of this new style of brick, adds another important advantage, to the long list offered by co-operative methods. In color, thickness, sanitary shapes, variety of designs, fire-proof qualities, polished smoothness and durability, these bricks recommend themselves to the favor of the general public, wherever they go. Without any effort in the line of advertising, the general demand for them has continued to increase, until brick-making has become the leading lucrative industry on the farm.

Among the new buildings at Solaris, most worthy of mention, are the theatre, and the two large school buildings, on either side of it. These structures, are by far the finest ones in the village. The affectionate pride they excite in the hearts of the villagers, is well deserved. Centrally located, on the east side of the public square, this triumvirate of noble buildings, claims the admiration of the beholder, from any point of view on the open square. The front walls are beautifully ornamented, in harmony with an architectural design, which is considered by critics, as exceedingly artistic. Inside, they have been constructed, finished, fitted and furnished, in accordance with a design, that will afford to the villagers, the highest order of education and amusement.

The theatre is two hundred feet long, and seventy-five feet wide. The schools, are each one hundred and seventy-five feet in length, by forty feet in width. They are separated from the theatre, by twenty feet of space. A roomy covered way from the rear, connects them with that building. In construction, care has been taken, to secure perfect light and ventilation.

The school on the left, is for pupils who enter the primary, and the first, second and third, intermediate classes. The one on the right, is for students, who may be promoted to the first, second and third, high schools. The seating capacity of each one, is ample for three hundred children. The decorations of the walls and ceilings are, to a remarkable degree, both educative and ornamental. The equipment of school furniture, such as seats, desks, dictionaries, text books, globes and outline maps; drawing-boards, blackboards and laboratory outfit; glass cases, for collections of geological specimens and minerals; life size, physiology models and charts; together, with a complete series of charts for the other sciences; is the best that could be designed or procured.

The theatre, is a very important part of the educative system. Fortunately, the accoustic properties, are remarkably fine! The entire interior, including the high ceiling, is decorated with such boldly beautiful designs, that they never fail to gratify the artistic sense of the beholder. At night, the charming effect of these embellishments, is intensified, by the use of a great number of brilliantly colored electric lights; which are skillfully grouped and interwoven, as a part of the general decorative plan. The wide seats, are designed for ease and comfort. They are richly and durably upholstered, with dark-brown, polished leather. The seating capacity of this cosey little theatre, is twenty-five hundred.

The colonists have found this histrionic temple, very useful It is an ideal place for farm and village festivals; and for all kinds of entertainments; such as orations, school exhibitions, graduation exercises, vocal and instrumental concerts and dramas; lectures, operas and every class of theatricals. It is also, equally useful and fitting, for stereopticon and biograph exhibits, of the astronomy, geology, botany, natural history, microscopical, and photographic clubs.

The large, well equipped stage and dressing rooms, offer a permanent, desirable home, for the musical, choral and dramatic clubs. At intervals of three months, four weeks in each year; excellent professional troups occupy the stage; presenting a fine variety, of wholesome dramas and operas. In this way, the stage of this farm theatre, is made to represent and reflect, the passing progress of the dramatic and operatic world. During the intervals between these star-company weeks, the home-talent club, presents regular, tri-weekly performances, under the supervision of a skillful director. The remaining nights are as a rule, pretty well utilized by the numerous local entertainments, before mentioned.

This brief sketch of the generous provision, made for the education and amusement of the people of Solaris, will, in connection with the nursery and kindergarten, hereafter to be described, show what the co-operative farm can do, when it undertakes to give to its people a class of educational training and amusement, which in many respects, is superior to the best that money can buy for the wealthy. It will also demonstrate, what can be accomplished, when the farm determines to produce, and to fittingly educate and train, a superior class of children, as the most important part of the legitimate work of a co-operative farm. The highest expression of agriculture! The culture of children as a fine art! The production of such children, as will make ideal citizens for a perfect Republic!

The practical class in farm chemistry, only twelve

in number, is an organization made up by a careful selection from the brightest minds and best thinkers in the colony. Under the leadership of Fillmore Flagg, it has accomplished some excellent experimental work; It has been able to add several valuable allied industries to the resources of the farm, in addition to those already described.

In breaking ground for opening the new mica and zinc mines, a great quantity of peculiar clay was discovered. This clay was of a very fine quality, entirely free from sand, gravel or other impurities. Yet, strangely enough, it would not make good china, porcelain, or pottery! There was a greasy smoothness of feeling possessed by this clay, which suggested its name, tallow clay. After considerable exposure to the air, it would crack and slack until finally dissolved into a fine powder. The class was puzzled. The members were on their mettle! The more they worked with this curious clay and failed, the more they became interested and determined to persevere, until some discovery should reward them. The greasy quality of the clay, suggested soap-stone. Now, the class members had long wished for some material out of which they could manufacture a first-class quality of artificial soap-stone. This tallow clay promised good results, if they could only eliminate the few constituents, which were not present in the real soap-stone. The weeks of careful research spent in this eliminating process, finally crowned the efforts of the class with a complete success. The result, was an artificial soap-stone of excellent quality. Even, when molded in thin plates, it would withstand exposure to intense heat for long periods of time, without warping or shrinking. It soon became evident, that it could be made more useful and more valuable, than real soap-stone.

After some weeks of experimental work, in various processes of manufacture, the right method was reached. Fillmore Flagg was convinced, that thousands of tons of this product, yielding a large profit, could be placed on the market much cheaper than the best quality of fire brick. For a great number of uses in the industrial arts, and for chemical furnaces, ore-roasting ovens, furnace linings, stove linings and even stoves, it would prove immeasurably superior. The popular demand for this new soap-stone, soon sustained the judgment of Fillmore Flagg. This demand continued to increase until the new industry, became one of the most profitable on the farm.

After the first success, the class in farm chemistry, in search of another prize, returned with renewed vigor, to attack the tallow clay. In working over the formidable heap of tailings, which had accumulated from the soap-stone experiments, the second prize was quickly found. It proved even more important than the first! This mass of rejected clay was found to be exceedingly rich in aluminum. Better still! It was just in the proper condition, to be most cheaply and easily extracted! It was a great find! The class members were crowned with laurels! Of course, they were jubilant. But they were not puffed up with pride! That, was not their style!

During the fifth year of the reign of the co-operative farm at Solaris, the following mining industries, were added to its resources. Valuable mines of mica, lead and zinc, were opened and successfully worked. Electric car lines, connected these mines with the freight depot at Solaris Station. There, the lead and zinc, high grade ores, found a ready market at good prices. The mica was prepared for use at Solaris. It was then sold at a fine profit, in connection with orders for soap-stone.

For two years, the canning factory, had furnished another avenue for profitably marketing large crops of sweet-corn, green peas; asparagus, tomatoes, peaches, and many kinds of perishable fruits and berries.

The demand for Solaris Vegetable Concentrates, and for Solaris Mixture Concentrates, has more than doubled. The same is true of the Solaris breakfast foods, and of the material for delicious breakfast dishes, prepared from mixtures of parched, sweet, and pop-corn.

The vineyards and the quince, peach, plum and cherry orchards, have reached the stage of full bearing. Improved methods, careful culture and the constant use of better chemical agents, for the destruction of insect enemies, have made the heavy crops of fruits from these vineyards and orchards, even more desirable and more salable than ever before. The farm income from grapes and quinces alone amounting to over one hundred thousand dollars per annum.

The quantity of jellies, jams, preserves and marmalades, made from small fruits, has more than doubled. The excellence of quality, and established reputation for absolute purity, has rapidly increased the demand for them at fancy prices.

Altogether, the rapid and continuous growth of the farm income, from its allied agricultural and manufacturing industries, has largely increased the wages of the co-operators. The purchases at the store have been correspondingly augmented. The sale of goods by the store, to surrounding communities, has been greatly extended. The result has been a constantly increasing volume of the seven and one-half per cent profits, steadily pouring into the insurance fund. Both the general service fund and the fund for purposes of education and amusement, have been equally benefitted. Fifty thousand dollars, have been added to the stock of goods, in the store. The store building, has been enlarged and improved. A large hotel for the accommodation of the constantly increasing number of visitors, has been erected and equipped. At all times, plenty of money has been at hand, with which to push forward all necessary farm or village improvements. The fame of such general prosperity, has gone abroad, in the land; placing the financial standing of the Solaris Farm Company, on a firm basis with the commercial world.

Five years of co-operative work, have convinced the people of Solaris, that successful agriculture, demands the determined effort, the best thought, the scientific work and the combined energy of a well organized force of earnest, unselfish, steadfast workers. They are very enthusiastic over the wondorful results achieved. Freed from the shackles and sins of a selfish life, they bear the unmistakable stamp of progress, socially, industrially, intellectually and ethically. Having cast aside the burden of care and worry about the future, both for themselves and their children, they have had a chance to grow and expand in the real sunshine of life. They have become dignified, self-poised, well dressed, educated, refined, cultured and polished men and women. Good citizens, of which, any commonwealth might well be proud! Vitally, and vastly more important! They have become dominant thinkers, who are capable of wisely and unselfishly, thinking and planning for the benefit of the Republic!

In the remarkable success achieved by Solaris Farm, our hero, Fillmore Flagg, has realized his highest ambition, his brightest hopes. Relieved from further responsibility, as general manager, by the last annual election of the Solaris Farm Company, he has had an opportunity to turn his attention to organizing companies, for the eight remaining farm sites. In this work, he has had valuable assistance from the officers and members of the company. With a view of making Solaris the present headquarters of the general movement; acting on advice of Fillmore Flagg, the Solaris Farm Company, has amended its charter, to increase the membership of the company to one thousand; doubling the capital stock. Five thousand acres of adjoining lands have been secured, the farmers from whom they were purchased, coming into the company as stock-holders. This course seemed necessary and wise, in order to properly balance the growing industrial and commercial importance of Solaris. With such a large increase in the number of co-operators, a surplus of capable young men and women, would be available, from which to select volunteers, as the nucleus of a corps of experienced officers for the newly organized farm companies. In this way, Solaris, as the parent farm, would become very important as the training school, for teachers that were to supply the wants of such new farms as might grow out of the general movement.

CHAPTER XXXVIII.

THE KINDERGARTEN AT SOLARIS.

AMONG the important buildings at Solaris, we must consider the large, well appointed nursery, kindergarten and mothers club combined. The mothers club occupying a handsome wing to the main building. Located just in the rear of the long row of palace homes, and connected with them by a long, wide, many-windowed hall, it has proved admirably adapted to the purpose for which it was built. This beautiful structure, is environed by a lovely lawn, charmingly variegated with flowers and shrubbery It is surrounded on three sides, by a wide, low veranda, only one step above the lawn. This veranda, except where a broad step connects it with the lawn, is shut in by a tall balustrade. By this means unguarded children are prevented from falling. A broad, overhanging roof, of picturesque design, covers the entire building. From the interior, many windows coming down to the floor, open on to the veranda.

The entire floor space, the full size of the main building, sixty by two hundred feet, is unobstructed by a partition. That portion devoted to the nursery, is only separated from the kindergarten by a low balustrade. A large skylight, in the central roof, floods this extraordinary room with an abundance of light. Screens of thin, white, silky cloth are so arranged, that this light may be regulated and softened to any desired extent. The lofty ceiling is arched, groined and decorated, very like a cathedral. The high walls are modestly tinted a pale green. A broad, beautifully designed, exquisitely colored border, in perfect harmony with the splendor of the ceiling, runs uniformly around the upper walls of this delightful room, adding immensely to the general artistic effect.

One peculiarity in connection with the floor, marks a wide departure from the ordinary arrangements of a nursery or kindergarten school. Six feet distant from the washboard, a depressed railway track, equipped with long platform cars, ten feet in width, having their surfaces just level with the main floor, describes a circuit of the room. Except at the places of entrance or exit, this circular train or section of floor on wheels, is guarded on either side by a low railing. These railings also extend across the cars, far enough from the ends to allow a four foot passage between each one. In material and finish, the floor of the train is uniform with that of the room. The railings are all of polished oak. Two cute little gates on each car open to the passage way at the ends.

The machinery which propels this exaggerated perambulator, is run by electric power. It is so adjusted, as to be perfectly under the control of the nurses and teachers in charge of the room. The iron frames from which fifty swinging cribs are hung, occupy considerable space on several cars. These cribs are for the exclusive use of infants, too young or too weak to sit up. The remaining space on the cars of this infantile merry-go-round, which the mothers' club members have named the Cargosita, is furnished with a remarkable variety of single and double seats, made low enough to be comfortable for children from eight to thirty months old. These seats are as artistic as they are unique! They represent on a small scale, ostriches, swans, geese, dogs, goats, horses, mules, zebras, camels, elephants, tigers, and lions; wagons, phaetons, cycles, cars and a great variety of pleasure boats. The seating capacity of the cargosita is about three hundred, the number of children in the nursery and kindergarten, who are under four years of age. Older children become inmates of the regular schools.

The cargosita, when ornamented with a profusion of silk flags, resplendent with gaily colored ribbon streamers, handsome mats and a choice collection of small potted plants, palms and flowers; becomes a thing of beauty, well calculated to capture and fascinate the childish heart. When the train is in motion, gaily spinning around this five-hundred-foot oval; the cribs and seats filled with bright happy children, smiling and crowing, their chubby little hands clapping in unison with the measure of such exquisite music as is discoursed by a giant orchestrion, or the electric piano, the vision becomes the lovliest and most inspiring one of a life time!

When we consider the cargosita as an instrument for education, we find that it is even more potent as such, than as a thing for amusement. For the purpose of educating the senses, thus laying a sure foundation, for a broad, healthy, harmonious, development of the mind, it is invaluable!

A child is the repository of infinite possibilities! Education, is the process of unfolding these possibilities, in harmony with natural law. To discover, and to apply this law, is the important work of the educator!

To Prof. Elmer Gates, and to his remarkable discoveries in Psychology and Psychurgy, the modern educator owes a heavy debt of gratitude! From the teachings of Prof. Gates, we deduce; that in brain building, that primary step in education, psychologic functioning creates organic structure, and that organic structure is a manifestation in the concrete, of the activities of the mind. In other words, that planted, watered and nourished, by the emotions of the individual, the thoughts, ideas, concepts and images which arise, create a corresponding growth of cell structure in the brain. That these brain cells become the working tools of the mind.

It follows then, that we cannot have thoughts, without first having sensations to form images and concepts, the soil out of which all thoughts naturally grow. Therefore, if in a practical way, all possibilities in the way of sensations, which may come through the avenue of each one of the child's senses, are fully developed; a sure foundation has been laid, for the largest possible development of brain and the corresponding growth of thought.

In the natural order of the growth of thought, nature prescribes the following sequence: A union of sensations, produces images; a grouping of images, produces concepts; a relationing of concepts, produces ideas; a generalizing of ideas, produces thoughts of the first order; a generalization of thoughts of the first order, produces thoughts of the second order: a still wider generalization of thoughts of the second order, produces thoughts of the third order; progressing in like manner, to the highest ladder of the mental scale.

In considering this order, we observe that sensations, form the base of the educational pyramid. All knowledge which comes to the ego, the seat of consciousness, must come through sensations produced by contact with material things in the domain of nature. Hence, as a primary step in educational work, a careful training of the senses, becomes a matter of the greatest importance. This training cannot be commenced, without first ascertaining what these senses are, and the natural order of their evolution.

Commencing with the lowest, we have muscle feelings, or the sense of musculation; the sense of touch, the sense of pressure, the sense of warmth, the sense of cold, the sense of smell, the sense of taste, the sense of hearing and the sense of seeing. Altogether, we have nine important avenues, through which the inner man may gain a correct knowledge of the outer world.

Professor Gates has discovered a system of sense training, which may be successfully applied to kindergarten children. In application, only a few minutes daily practice by each child, is required. By this training, in extending the upper and lower thresholds of sensation, the capacity of each sense, may be doubled from five to eight times. To the inexperienced, this proposition is so stupendous, that it seems almost unthinkable! However, we may state parenthetically, that an application of this system, to children in the Solaris kindergarten, has shown such marvelous results, that its efficacy and excellence have been well established. It has proved fully equal to the demands of twentieth century progress!

Turning again to the teachings of Prof. Gates, we learn that mind is the key-stone and the arch of life, the all-containing attribute, which combines all forms of its expression: that to properly cultivate the mind, is to extend the scope and usefulness of life. Hence, that in choosing a system of education, which will be in harmony with planetary evolution, therefore, the easiest and most natural. We must never lose sight of one great, central, primal fact. It is this. The mind of the child, which is to be unfolded, is the production of the cosmic universe; therefore, cannot be in fundamental antagonism with it. It follows, then, that if children gather their sensations, images, concepts, ideas, and thoughts, directly from the phenomena of that universe, they will acquire a kind of knowledge, so real, so superior, that it will stand the test of an eternity. It is actual knowledge! There is no theory, no speculation, no guesswork about it!

The sciences, are facts regarding the phenomena of the universe, classified and arranged in an orderly manner. All facts of every kind, naturally fall into the domain of some one of the sciences.

Man, as the highest expression of the planet, in his three-fold nature, becomes the gleaner, the classifier, and the repository of these facts. A beautiful exposition of the clever handiwork, of the law of action and re-action. As a cosmic unit of the larger cosmos, the more perfect his knowledge of the universe, the more complete, is his store of knowledge in relation to himself.

Children, in order to become properly equipped students, must, when ready to take up the sciences, be prepared to determine what the actual sensations are, out of which the different possible images of the sciences are composed. To achieve the most thorough education possible, they must know the actual number of concepts in each science, and precisely the images out of which they have arisen! They will then be prepared, to collect and classify, the mentative data of the sciences. That is, they will be able to determine for themselves, experimentally, the sensations, images, concepts, ideas and thoughts, which belong to each one.

Practice in this useful training, will lead the pupil, to the higher, wider generalizations of thought, which belong to the domain of pure reason. In the work of classification, by detecting differences, a knowledge of the inductive process is gained. Similarly, by detecting likenesses, a knowledge of deductive reasoning is acquired.

The body, like the brain, being composed of a cooperative colony of more or less intelligent cells, is an important part of the mind, which responds to educational training. True education, then is a development of both mind and body, in accord with the law of natural evolution, that embraces all there is in the domain of morals, pertaining to right thinking, right living and right doing. In other words, the action of the mind comprehends the physical, intellectual, moral and spiritual expression of the individual. Therefore, by the rightly conducted processes of a higher education, we may form an evenly developed character of the highest order. A character, unfolded physically, intellectually and spiritually, in harmony with the requirements of cosmic law. Hence, the imperative necessity, in the early training of children, of introducing the first steps of this system of true education.

From these premises we must conclude, that the first four years of a child's life, should be devoted to some systematic method, for acquiring a most complete equipment of exact images, which will afford the basis for typical sensations, emotions, ideas and thoughts, regarding things in the domain of nature, about which, later in life, the child must know in order to become educated. To this end, children must have opportunities during these important years of image building, to experience all the sensations, and to form all the true images, that can come to them through the senses of seeing, hearing, tasting, smelling, touching, feeling and sensations of temperature, such as heat and cold.

It is of the utmost importance, that these early images, which are to become the standard of the mind, in all judgments of future years; should be made as complete and as perfect as possible.

A child is primarily and instinctively imitative. From the first dawn of intelligence, children strive to emulate the acts of their brighter, older and better-taught associates. Hence, the necessity for a nursery and kindergarten training, such as the one instituted at Solaris. Practical work, in this novel and magnificently equipped institution, has proved conclusively, that, even in early infancy, associated together in happy groups, children acquire intellectual, moral and physical training, much more easily and swiftly, than is possible under any other circumstances. This affords another demonstration, of the efficacy of co-operative group work, in the primary steps of education.

The cargosita, is well calculated to offer children the most perfect conditions, for accumulating a well selected store of sensations and images, through the avenues of the different senses. A teacher or nurse, usually some member of the mothers' club, is seated on each car as the center of its group. It becomes her pleasure, to direct attention to the various objects. Let us follow the cargosita with its precious freight, as it slowly moves around the oval.

Images produced by the sense of seeing, are first

in order. Large sheets of thick, heavy paper mounted on cloth, seven in number, displaying the different colors of the rainbow, are hung at uniform intervals around the room. They can be raised or lowered, to reach an easy angle of vision from the cars. After each primary color, appear half-width sheets of the same height, displaying the various hues, tints and shades of that particular color. Printed across each sheet in large white letters, is the name of the color, hue, tint or shade. Altogether, this color scheme forms a combination of great length, of such remarkable variety, that it becomes for the little ones, a well nigh inexhaustable source of fascinating amusement.

Red, with its various hues, tints and shades, is the first color to be exhibited. Three days later, another color series is substituted. This course is continued until the entire series is finished. The children have experienced in a regular sequence, the sensations and images, produced by the entire scale of color. These mental pictures have been repeated so often, in connection with the muscular sense of exhilarating motion, that they have become permanently enregistered in brain-cell formation. A review every few months, serves to fix these images more firmly in the brain.

This primary course of educative work is continued, by taking up consecutively, in regular order; on a separate series of sheets, life size, naturally colored photographs, of fishes, reptiles, insects, birds, animals, and people. Later, geological specimens, glass, rocks and minerals. To be followed by pictures of life in the vegetable kingdom, flowers, fruits, plants and trees. Again, with photographs of works of art, paintings and statuary.

Interspersed with this general course, are short lessons, offered to produce true images, in the hearing, smelling and tasting areas of the brain.

First, by repeating at different times, while the cargosita is in motion, with its cargo of infantile passengers, all of the best musical compositions, executed vocally, and on the electric piano, the giant orchestrion, the violin, and a great variety of other musical instruments. These lessons in hearing, are repeated and varied, until the children have become familiar with most of the sounds in the tone scale. The mental sound images produced, have been associated with the happy scenes of this merry kindergarten life. By this interweaving of pleasant sensations, they have become more firmly fixed in a healthy group of brain cells, thus planted and established in the hearing areas of the brain.

Second: In a similar manner, the taste sensations and images, are produced and registered. •Day after day, one by one, tiny packages of confections, beautifully wrapped in brilliantly colored papers, are given to the children while on their cargosita excursions. These interesting lessons are continued, until the entire range of savors has been exhausted. The curiosity, excitement, pleasure and eagerness exhibited by children, in these tasting investigations, is something surprising.

Third: Flowers, beautiful flowers of all kinds, are largely used in producing sensations and images, to be registered in the brain areas of the sense of smell. The essence of odors which cannot be gotten from flowers, are used to saturate small sachet bags, of charming color and artistic design. These bags make attractive play-things for the children. While using them they soon, unconsciously, become very skillful in detecting the slightest differences between the various odors. Brain areas usually left barren, are now filled and developed.

Later in life, when children come to study the different sciencies, this ability to detect the presence of the slightest odor, becomes invaluable, in the difficult work of classification. With such an unusual equipment, they will be far in advance of those pupils, who have not wisely, left uncultivated this important sense of smelling.

In connection with the regular course of exercises, prescribed for third and fourth-year children, there is introduced in the play and work rooms of the kindergarten, a special training, designed to develop the various sensations of heat and cold: changes in temperature, from one extreme to the other: sensitiveness to touch: to recognize any degree of pressure, from zero to the violence of pain: ability to detect size, length, breadth, and thickness: degrees of smoothness, elasticity, and hardness: all through the senses of touch, pressure, and muscular feeling.

Interesting plays are invented for the children, into which, these exercises are skillfully introduced. These plays, have a peculiar fascination. They excite an intense interest, which seems to always attract and hold the child's attention, until there is enregistered, in regular sequence, in the touch areas of the brain, all the sensations and images, which can be produced by many weeks of training, in this systematic course.

The training of the senses, is also carried forward through the medium of such plays as are calculated to bring out the child's capacity to distinguish the least noticeable difference, in pitches of color, degrees of light, pitches of sound, with its degrees of volume and loudness; together, with ability to discover the least noticeable difference, in resistance to pressure, or the slightest increase or decrease of rythmical motion, etc. The lines of least noticeable difference, in the capacity of the various senses, having been well established, the training commences along those lines. Very soon, in the brain areas of the senses under training, there comes an increased cell growth, which gives added sharpness and capacity. The line of least noticeable difference, is moved one step nearer the limit. This process is continued with each sense separately, until the limit for all has been reached. As a general result of this training, we find that the child has acquired an extraordinary reinforcement of brain power and intellectual acuteness.

Regular kindergarten work, for children at Solaris, between two and four years of age; is again reinforced, by adding to the list of exercises, a large number of plays, which introduce the variously colored, lettered blocks, so successfully used in Fern Fenwick's early training, during her seven years of Alaska life.

The collection of blocks, is a very large one. It is calculated to furnish a series of new combinations, which cannot be exhausted, in the plays of one whole year. These blocks are made and colored with the greatest care. The groups or families, are distinguished, by size, shape and color. The Alphabet blocks, are large cubes, painted white, with the letter showing in black on every side. All other blocks, have a uniform thickness of one-half inch. They are as large as can be fashioned from blocks two inches square. The names appear in white letters, on all alike.

The astronomy blocks are star shaped, painted blue. The geology blocks are diamond shaped, painted brown. The chemistry blocks are hexagonal in shape, painted red. The geography blocks are globular in shape, painted gray. The blocks representing physics, are octagon shaped, painted yellow. The botany blocks are oblong, painted green. The physiology blocks are triangular in shape, painted pink. The history blocks are square, painted black. A large number of the key-words of the sciences, are painted on blocks, which, in size, shape and color, are counterparts of those that represent the heads of families to which they belong.

This scheme of blocks, furnishes the ground work for the construction of a great number of games, for the amusement and edification of the children. Games of word-building, such as spelling out the names of fishes, insects, reptiles, birds and animals. Also of building the names of familiar things, houses, stables, light-houses, factories and mills; rivers, ponds, lakes, mountains, trees and fields; hats, shoes, coats, cloaks and other articles of clothing; common household utensils in every day use, such as pots, kettles, pans, pails, cups, knives, forks and spoons; stove, shovel, tongs, mop and broom; toys, dolls, balls, kites, tops, etc.

By the use of many such ingenious games, the children unconsciously become familiar with the names of the sciences, and with all the principal words, which belong to each one. For example: Names of heavenly bodies in the domain of astronomy. The sun, the moon, the milky way, the planets, the constellations, the polar star, and the names of twenty stars of the greatest magnitude: In the domain of geology, fossils, shells, minerals, rocks, shales, clays, gravels, and the names of geological periods: In the domain of chemistry, the names of acids, gases, metals, crucibles, retorts, mortars, and the names of a great variety of chemical combinations: In the domain of geography, globes, hemispheres, continents, islands, oceans, gulfs, bays, and straits; equator, tropics, circles, longitude, latitude, etc. These examples, will furnish an approximate idea of the wide scope in scientific names, covered by these key-words, when applied to all of the sciences.

In such plays of science grouping, the interest and pleasure of the children is intensified, by applying a system of personification, to the families of the different sciences: For instance, Mr. Astronomy Blue; Mrs. Geology Brown; Mr. Chemistry Red; Mrs. Geography Gray, etc.

In the greatest and most useful of all games, the game of classification: Groups of children, spend hours with their teachers or directors, in separating and classifying, heaps of miscellaneous blocks, bearing the names of the sciences and the key-words belonging thereto. They are silent, absorbed, contented, thoroughly interested and happy So intense is the interest displayed, that after the fourth or fifth game, every child can correctly classify the blocks, by quickly placing them in the groups to which they belong. They rapidly learn to call the name at sight, which is printed on any block they may happen to pick up. Those who have not learned to read by playing word-building games with the alphabet blocks, only need to have an unfamiliar name, repeated to them three or four times by the director, and it is fixed. Size, shape and color of block, with length of name and shape of its letters, soon serves to make the little ones, perfect masters of the most difficult names.

These children have learned the value of time. They have learned to appreciate the joyousness of useful amusement. They have no desire to clog their minds, with the untruthful trash of fairy tales and Mother Goose stories, which played such an important part in nineteenth century methods. They no longer need such silly things, as a source of amusement. They seem to realize, that they only have mind-room, for the truthful, the useful and the practical.

The value and significance of figures, is taught by the game of forming the pyramid. On badges of broad, blue ribbon, are printed large gold figures, from one to ten. Inside the oval, in the center of the large room, ten rows of seats are arranged : with one seat in the first, and ten in the last row. That is, one seat is added to each succeeding row.

At the commencement of the game, when number one is called by the director, the little boy or girl, who is decorated with the badge bearing that number, takes the first seat, which forms the apex of the pyramid. The two children who wear number two badges; when called take seats in the second row. Observing this order, the calling is continued until the seats are filled, and the pyramid of fifty-five children is complete.

The director, having taken a position a short distance in front of the apex of the pyramid, proceeds to call the children to their feet. Calling by number, commencing with the tens, the rows rise in succession, from the base to the apex. Each row is called upon to perform some part of a short series of graceful gymnastics. Then, the whole group in unison. Later, these exercises are made more interesting, by giving each child a small silk flag. In this part of the game, the children are at their best. The picture they make, is just lovely!

In the closing part of the game, the children are seated and the mathematical exercises are introduced. The director says: "Each child has one nose. How many noses, have the number tens? Again, each child has one body. How many bodies, have the number nines? Each child has two eyes. How many eyes, have the number eights? Each child has two ears. How many ears, have the number sevens? Each child has one mouth. How many mouths, have the number sixes? Each child has two arms. How many arms, have the number fives? Each child has two hands. How many hands, have the number fours? Each child has two legs. How many legs, have the number threes? Each child has two feet. How many feet, have the number twos? Each child has ten fingers and ten toes. How many fingers and toes, has number one?" These questions are varied and repeated, day after day, until every child in the pyramid, can answer any one of the questions, correctly and promptly. To be chosen as a member of this game, is a coveted honor, it is conferred as a reward for good conduct. Consequently, the pride and pleasure exhibited by these decorated and selected children, is commensurate with the importance of this very primitive class in mathematics and physiology.

This very brief outline, of the plays, exercises and studies, which form the nursery and kindergarten course, for children at Solaris, who are under four years of age, will serve to show how much important knowledge, a child can accumulate during those fruitful image-bearing years, while pleasantly and zealously engaged, day after day, in a series of wisely directed games.

In playing these games, the children have become interested in, and have learned a very large number of useful words. These words in the mind of the child, are as familiar and as easily remembered, as are the names of favorite toys, such as balls, bats, kites and dolls. This wide vocabulary of key-words which has become the mental property of the child, has planted in the mind the necessary images, which in future years of study, will serve as a sure foundation, for the quick and easy mastery of all branches of useful knowledge. Many a man of the world has gone through life, without acquiring such a vocabulary.

Considering this primary course of study from another point of view, we have an illustration of the value of a method for cultivating the faculty of memory, which differs widely from any thing known to ordinary systems of education. From this illustration, we perceive that the perfectness and permanency of memory, is dependent on the foundations which have been laid for it, by the quantity and quality of sensations and images, regarding the things to be remembered, which have been registered or planted in brain-cell formation. These living images, fixed on the sensitive plate of the brain by the law of vibration, in a manner somewhat analagous to etching on the cylinders of a phonograph, are capable of being reproduced by the will-force of the individual. From these premises, we have gained a new definition for the word memory. It is a process of refunctioning or reregistering, any sensation, image, concept, idea, or thought, which at any time has become a part of the growth of the brain.

In the child's mind, memories regarding objects or words which have become familiar, are as a rule, closely connected with memories of keen enjoyment, resulting from participation in some childish sport. These memories are many times repeated. A few small groups of brain cells have become dominant in growth, because they have received the full force of the entire stimulating power of the brain. Hence, the memories of childhood, are much more enduring than those of after life. Hence, it becomes a matter of the utmost importance, that these early images, should be connected with the greatest possible number of natural objects, their names, and the key-words of the sciences, which are used to describe them.

In these restless years for the little ones, it becomes a matter of great moment, to keep their minds busily employed, at what appeals to their self-consciousness, as some useful work. In this respect, the popular science games, gratify and completely satisfy the pride and dignity of these embryo men and women. The mind is naturally unfolded, The brain areas, are all evenly and harmoniously developed. The children, when so usefully employed, are kept amiable. They do not become nervous, irritable, cross, or vicious. They are taught, as soon as they can walk and talk, that the self-respect and innate dignity, which belongs to them as little men and little women, demands that they should always treat each other lovingly, politely, kindly, unselfishly. It is continually urged upon them, that they must learn to obey the nurse or teacher, without delay, without a murmur; that they must not cry or be fretful; that in these things, they must always strive to imitate the good acts of older comrades or playmates. In this way, the moral unfoldment and education of the child, keeps pace with the intellectual and the physical. Altogether, the effect is most excellent! Thousands of children have gone to ruin, for the want of just such training, in the first four years of life!

The planning and final organization, of this novel scheme for nursery and kindergarten training, has been the joint work of Fern Fenwick, Fillmore Flagg, Gertrude and George Gerrish. In striving for the best results, this quartet of co-operative educators, have been ambitious to perfect a system, which would satisfy the demand for a natural, harmonious unfoldment of the well-born babies, which were to represent the highest product of Solaris Farm.

The success which has attended the practical operation of the scheme, has made them very happy. Towards this success, Fern Fenwick has been able to contribute largely, on account of her early Alaska training, and her thorough knowledge of the improved methods, growing out of the important discoveries made by Prof. Gates.

In applying the system to the class work of the regular schools, the long experience, trained skill and natural aptitude as teachers, of George and Gertrude Gerrish, has proved wonderfully effective.

By supplementing the system, with a very complete course of manual training in the use of tools, and in acquiring a competent knowledge of the industrial arts, Fillmore Flagg has been equally successful, in educating the muscular children, and in arming them most effectively, both mentally and physically, for the practical work of life.

Altogether, the complete course, results in an all-round development of brain power, more than five times greater than that offered by any other system. A result, which marks the beginning of a new educational era. A result, which promises to give to the world, a dominant race of thinkers, whose ability to bless mankind, is to be so great, that it cannot now be estimated.

CHAPTER XXXIX.

AN UNEXPECTED VISITOR.

In the month of August, 1911, six years after our first introduction to him, we find our hero, Fillmore Flagg, seated in his private office at Solaris. This office was located in a building on the public square, near the store, which has been especially designed and constructed, for use as the central office for the general co-operative, farm movement. Here, Fillmore Flagg, has been busily engaged for more than two months, in planning the preliminary work for eight new farms. For the moment, he seems absorbed in a dreamy reverie. From this, he was sharply aroused by the entrance of a messenger, who announced a visitor. The visitor proved to be none other, than our old acquaintance, George Gaylord. The greetings, exchanged between these re-united college chums, were cordial indeed! In the conversations which are to follow, the reader will find a continuation of the story of Solaris Farm.

"Shades of venus! How well you are looking,

Fillmore! I need not ask how you have fared since last we met! One look at your face, tells the whole story! The goddess of good fortune, must have smiled on you right royally! I congratulate you most heartily! The fame of your exploits here at Solaris, has reached New England! What a lovely village you have made! And the farm too, is just delightful! To behold it, is well worth the price of a long journey! Of course, at some convenient time, you are to show me the farm, and tell me all about it."

"Thank you George, for your congratulations; You have surmised correctly! I have been prospered, far beyond my most sanguine expectations! At the proper time, I shall take pleasure in relating the whole story for your benefit. Now, I am anxious to hear something regarding yourself. Tell me, my dear fellow! To what piece of good fortune, do I owe this unexpected visit? And, may I hope, that the goddess you just mentioned, has been equally gracious with her smiles for you!"

"It is a long story, Fillmore, and I can assure you it is not a pleasant one. It seems a pity to mar your peace of mind by relating such a miserable tale of woe! During the past five years, the unkind fates have frowned upon me, and I have suffered much! In order to give you an intelligent reason for my visit to Solaris, I must tell you of some good, and many bitter things which have transpired, since we parted at the hotel on Mount Meenahga."

"Really! George, I am sorry for your misfortunes! But surmising so much from your preparatory statement, I now wish to know all that you can consistently tell me. For the bitterness and suffering, you have my sympathy in advance."

"Thank you Fillmore! I knew that I could rely on your sympathy and friendship, under all circumstances. Please pardon any lack of coherence or orderly arrangement of details, in what I am about to relate.

"Late in the month of November, which followed our parting in the mountains, in accordance with previous arrangements, I took charge of the church in the New England city, where my uncle George resided. My relations with the members of the congregation, proved as pleasant as could be desired. I became acquainted with Martha Merritt, my uncle's niece by marriage. She was a beautiful girl! Very winning, sweet and amiable. I soon became fond of her company. This seemed to please both my uncle and my mother. I could see that they had set their hearts on a marriage between Martha and myself.

"About the middle of the following January, acting on a suggestion from uncle George, I asked Martha for her hand in marriage. After taking a whole week for consideration, she finally consented and we were engaged. Some days later, I urged her to name an early day for our wedding. Very much to my surprise, she said 'You must not hurry me, George! You must give me time!' I hastened to assure her that I did not wish to be inconsiderate, and begged her to take another week, in which to fix the date. During this time, I saw very little of Martha. In the brief interviews that followed, she was pale and agitated. At the end of the week, again her old-time self, she came to me with the news that our wedding day had been fixed for the fifteenth of June, five months distant.

"Early in February, the clouds of disaster began

to gather. My mother was confined to her bed with what proved to be a serious illness. After four months of almost constant suffering, which she bore with the patience and fortitude of a martyr, she was borne across the dark water, to join that vast majority, that silent, mysterious, ever increasing host of the buried dead.

"My mother was buried on the fifteenth of June. Overwhelmed with grief, I readily assented to Martha's suggestion, that our wedding should be postponed until the first of October. Recovering slowly from the shock of my bereavement, I turned eagerly to Martha, for loving consolation. I was horrified, to find that her affection for me had turned to ill-concealed aversion! There was a terror-stricken, haunted look in her eyes, as she strove in every possible way, to avoid being left alone with me even for a moment, which frightened and almost crushed me with grief. I knew that something dreadful, must have happened! She was so pitiful to behold, that I could not be angry or jealous! But, I resolved to know the truth. At the first opportunity, I demanded an explanation. Bursting into tears, she told me the story of her bitter experience.

"Falling on her knees beside my chair, Martha implored me to be merciful. 'George,' she said, 'I know that I am the most wretched, and the most desperately wicked girl on the face of the earth! You have been so kind, and I have treated you so shamefully! How, can you ever forgive me? The only reparation that I can now make, is to tell you the whole truth, without reservation. Ten months before I saw you, while I was at school near Boston, I met Phillip Plato. The fates would have it, that we should fall desperately in love with each other, at our first meeting. In a short time we were engaged. In entering into this engagement, I did so without the knowledge of my uncle, or any friend. I did not stop for a moment, to consider my duty to uncle George, who had always been so good to me. I could think of no one but Phillip, and of my love for him. In the delirium of love's first dream, the weeks passed as days! Alas! The dream was passing brief! Somehow, Phillip's parents became aware of our engagement. They were very wealthy, and exceedingly ambitious to have Phillip marry more wealth. Angry with him, they came to me and cruelly declared, that they would never allow him to wed such a fortuneless girl! With look and gesture of scorn, they told me that they were just on the eve of going abroad, taking Phillip for two years of travel, in which they should strive to cure him completely of his insane infatuation. This, then was the end of my romance. My cruelly wounded pride, rose up in rebellion. I was furious! I returned scorn for scorn! I bade them begone!

" 'I returned to my uncle's home, my heart hot with the indignation of an outraged pride, and filled with a determination, to show to the world no sign, but to use all my strength of will, to cast Phillip out of my life; to utterly forget him and his selfish, greedy, heartless parents. When you came, George, I was more anxious than ever before, to please my uncle in every possible way. I foolishly imagined, that in encouraging your attentions as a lover, I was helping myself, to forget my love for Phillip. Oh! What a terrible, cruel mistake! How terrible, how cruel, I was soon to realize. You will remember, George, how strangely I behaved at that interview, in which you asked me to fix the day for our wedding.

Let me explain. A few hours previous, while I was lost in one of my occasional fits of melancholy moping, the voice of Phillip came to my ears with startling distinctness. The voice said Martha, you must remain true to me! I love you as devotedly as ever! I am determined, never to give you up! I am coming home to wed you! I am surely coming! Wait for me! These words kept ringing in my ears, like the tolling of a funeral bell. They thrilled me through and through! The barriers of my pride gave way. The returning tide of my love for Phillip, swept in upon me with such force, that my heart almost ceased to beat! I was faint, deadly faint! When I recovered consciousness and afterwards, at our interview, I was absolutely wretched! Your request, added to my anguish. I was powerless to answer, I could only beg for more time. All through that dreadful week, I strove to convince myself that my ears had deceived me, that the voice was not real, only a phasma, a hallucination, born of my fits of melancholy. Unfortunately, I finally succeeded!

"'Now, George, you shall hear the sequel, the climax of my wretchedness. The day before your mother died, I received a long letter from Phillip. It was written at Rome. Every line of that letter, was eloquent with Phillip's steadfast devotion, and love for me. In brief, a complete verification of what the warning voice had told me. His parents had relented. He was coming home to make me his bride. He had planned to arrive at Boston, in time to celebrate the New Year. He spoke of a long letter, which he had written to me, just on the eve of his going abroad. In that letter he had assured me of his undying love, of his determination never to give me up. In closing, he had begged me to wait for him, to remain true to him. He had repeated its contents, because he had been constantly haunted with the idea that the letter in question, had failed to reach me. And so it had.

"'This, George, is the summing up of my misery! It has filled my heart with the anguish of despair! I can never love anyone but Phillip! I cannot marry you, George! I cannot! It would be an unpardonable sin against you, against my own soul! What shall I do? What can I do? What atonement can I ever make, for the shame, the humiliation, the suffering, which I have brought into your life?'

"In this brief sketch, Fillmore, you have the substance of Martha's sad story. I believe it was absolutely true. I was deeply moved, by her abject misery and humiliation. A great wave of tender sympathy, swelled in my heart; blotting out all thoughts of self. I gave her back her engagement, and bade her go free; free to marry whomsoever her heart had chosen; assured of my forgiveness, and of my wish for her future happiness. I need not repeat her grateful thanks. From this time forward, our lives were widely separated.

"During the long tedious months that followed, I was going through a bitter, humiliating experience. I strove by every effort to so interest myself in my church work, that I might forget my griefs and my disappointments. In this, I failed utterly. I found to my amazement, that I did not possess a thorough belief or confidence, in the efficacy of the atonement, the very ground work of the entire scheme of christian salvation. Without this belief, I could not hope to do effective work in the ministry. No doubt, this was the cause of my lack of interest in my pastoral duties; the one thing, during this time of trials, which most disturbed my mental equilibrium, and added to the intensity of my sufferings. My growing antipathy towards all kinds of church work, daily increased the mental tension, caused by anxious seasons of watching, praying, and fighting, against the farther dominancy of this monstrous antipathy. All opposing efforts proved useless. With each succeeding week, my Sunday services became more burdensome, more perfunctory, more unsatisfactory, more self-accusing. At last, in self defense, the church trustees proposed my taking a year's vacation, for recuperation.

"This welcome respite, I gladly accepted. My vacation, is now nearly finished. I cannot go back to my church. I do not wish to go. I realize, that I am wholly unfitted for its duties. I feel, that I have made life a failure! In fact, Fillmore, you see before you in your friend George Gaylord, a man who is aimlessly drifting on the sea of life, like a ship without a rudder. A man not yet thirty, without a home, without ambition, hope or purpose! Possibly, I may be in the clutches of some approaching attack of nervous prostration, I hope not, I am sure!

"You must pardon my prolixity, Fillmore. I will now give you the reason for my present visit to Solaris. After my mother became very ill, some weeks before her death, she received a letter from Caroline Houghton, a life long friend, an old schoolmate. At that time, Mrs. Houghton was residing in a small town near Denver, Colorado. She was a widow with scant means of support; with only one child, a daughter. Mrs. Houghton, in her letter, said: 'I am dying among strangers! I am leaving my darling daughter alone in the world, without money, without relatives; simply in charge of recently acquired friends. As a last request, I beg you, after I am gone to exercise a protecting care over my orphaned child!'

"This letter worried my mother greatly. I think if she had been well, she would have hurried to Mrs. Houghton's bedside. After some delay, she finally turned the letter over to me to answer. Just at that time, my mind was wholly preoccupied with preparations for my fast approaching wedding day; and also, with the adjustment of a number of important church matters, which demanded my immediate attention. Without taking time to read the letter, without realizing its importance, or its urgency; I mechanically placed it in my desk, thinking meanwhile, that when the time came in which I could pen a reply, I would then confer with mother for further instructions. Unfortunately, the letter became misplaced and all memory of its existence, passed out of my mind!

"One month ago, while busily engaged in assorting and rearranging a confusing mass of papers, I found the lost letter. After reading it carefully, I became conscience-smitten, as I thought what serious results might have followed my criminal negligence. I then commenced a search for this young lady, which has finally lead me to Solaris. I have traced her here, as a member of your colony. Her name is Honora Eloise Houghton. Do you know her, Fillmore! Is she here?"

"Make yourself perfectly easy, friend Gaylord! She is here! She is all right! Miss Houghton does not need your protecting care, or the protecting care of anyone. She is abundantly able to take good care of herself and of plenty of other people besides! She can dissipate your troubles in a jiffy! She can give you something to think of, which will not fail to hold your close attention. She can soon find a work for you, in which you will be interested in spite of yourself! In fact George, Honora Eloise Houghton, is one of the brightest, most independent, capable, self-poised, self-supporting young women at Solaris! If she should kindly consent to take you under the brooding care of her protecting wing, in one month's time you would not know yourself, you would be transformed into a new man! But, Miss Houghton is a very busy woman. One of the most useful on the farm! Just at present, she is the leading director of the nursery and kindergarten school; the principal female teacher, in the gymnasium; the president of the dancing club; the secretary and treasurer of the physiology club; and vice-president of the botany, chemistry and history clubs. After faithfully performing the duties belonging to these offices, she still finds time to do a great amount of scientific research and reading; so much, that last year, she easily carried off the prize, which was awarded to the best qualified, scientific student among the young ladies at Solaris."

"Stop, Fillmore! You grieve and astonish me! You surely must be jesting, in dishing up this long rigmarole, about Miss Houghton's accomplishments! After what I have told you, I cannot conceive how you can fail to understand, that I am not in a mood for jesting. As for the girl, I very much desire to meet her, that I may have an opportunity to express the regrets and apologies for my unfortunate neglect of her mother's letter, to which she is so justly entitled. This painful duty once performed, my interest in Miss Houghton will cease."

"I assure you, George, I am not jesting! I am very much in earnest! I think I understand your case thoroughly. I know that you do not realize the seriousness of that paralyzing, apathetic condition, into which you have fallen. I do not think you need condolence, or any form of mild sympathetic treatment. I am sure you do need very much, to be aroused by new associations, scenes, friends and acquaintances; strong magnetic people, with ideas so radical, so startling, that by one quick wrench, your line of thought may be diverted into some entirely new channel. If therefore, in my talk to you about Miss Houghton, I have succeeded in arousing your indignation, in the slightest degree, I shall be encouraged by knowing that my efforts for your good, have been made in the right direction."

"Pardon me, Fillmore! I fear I have been hasty! And, that I have entirely misjudged your motive! I am now in a much better frame of mind, to listen attentively to what you have to say."

"That sounds much more reasonable, George. I will now return to my description of Miss Houghton, which was broken off by your interruption. For the reasons I have just stated, I believe that Miss Houghton, is the one individual in a thousand, whose acquaintance just at present, would prove most beneficial for you. Of course you have not seen her, you do not know her; therefore, you cannot appreciate the peculiar charm of her magnetic presence, or the force and dignity of her attractive character. For this reason, a personal description, will fail to give you an adequate idea of the noble type of womanhood which she represents.

"However, George, after these preliminary remarks, I hasten to assure you, that as a woman, Honora Eloise Houghton, is a goodly person to behold. One inch less than six feet in height, straight as an arrow, broad of shoulder, and round of limb, swift of hand and foot, lithe and willowy in every motion, her commanding figure possesses the grace and beauty, of a Venus and a Diana combined. Her large, full, well turned neck and throat, fittingly supports a symmetrical, well poised head, of the same noble proportions. A long, thick, luxuriant growth of golden hair, brilliant with changing hues of a coppery tinge, seemingly so surcharged with electro-magnetic force, as to form a halo of sunshine around both face and head, is her chief personal adornment. Her large, oval face, well formed mouth, strong white teeth, firm chin, finely arched, strongly defined brows, broad, smooth forehead, and straight grecian nose; all denote a character of marked type and unusual force. Full, clear, gray eyes, set well apart, beautifully and mirthfully expressive, together, with a bright, ruddy complexion, are both indicative of Miss Houghton's perfect health and strong, vital, nervous-sanguine temperament. With this temperament and such a magnificent physique, reinforced by wonderful psychic powers, she is an ideal healing medium. The very personification of health! Such is the potency of her magnetic force, that among the people of Solaris, cures performed by the simple process of laying on of hands, have made her the marvel of the village; they have won for her the confidence, respect, admiration and love, of every member of the colony; man, woman or child.

"In conclusion, George, I may say with pride, that Miss Houghton represents one of the noblest of women, which may be discovered, evolved or grown by the co-operative farm As an exponent of what the movement can do for woman, she is a shining example, of which our people may well be proud!

"Try to be patient with me, George! I have described this young lady, at such length, in order that you may meet her without prejudice. We will now go in search of Miss Houghton, for an interview. After introducing you, I will return here. When the interview is at an end, I will have my light, road mobile ready, and we will take a spin around the farm. Afterwards, if there should be time, we will take a run over to Fenwick, ten miles away."

"That arrangement will suit me very well, Fillmore! I am now quite curious to meet Miss Houghton. After my interview with her is concluded, I shall be delighted to accompany you on a mobile excursion over the farm. I have in mind a host of questions, which I wish to ask; after my tour of inspection, I am sure I can frame them more intelligently."

Four days later, we find George Gaylord, again seated in the office with Fillmore Flagg. They are speaking of things which have transpired, during the interval named.

"You are looking decidedly better, to-day, George! I congratulate you! After the fright you gave me, while at the club dance, that evening after your arrival at Solaris, I thought you were ticketed for a long, serious illness.'

"Really, Fillmore, I have Miss Houghton to thank for being able to again walk and talk with some degree of steadiness! She is truly, the most marvelous woman, that I have ever met! There seems to be a healing power in the very touch of her garments! I feel quite sure, that she has saved my life. I ought to apologize to the members of the dancing club, for the very awkward sensation, which must have followed my unfortunate collapse; that sudden attack of giddiness and loss of consciousness. Miss Houghton tells me, that the attack lasted over an hour, after I had been placed on a cot in the hospital. Were you there, Fillmore?"

"What a question, George! Of course I was there! That one hour, seemed three to me. Knowing something of your critical condition, I was blaming myself, for having foolishly attempted to crowd so much into your first day's experience at Solaris. However, Miss Houghton assured me, that I need not be alarmed over the trance-like condition, into which you had fallen. She seemed to understand your case from the first, and declared that she could cure you with a few days treatment. She further stated for my benefit, that I was in no wise responsible for the attack of vertigo, which in your condition, was liable to occur at any time.

"So far as the dancing club people are concerned, no apologies on your part are needed. They understand the circumstances, and wish me to assure you, that they will rejoice with you over your speedy recovery. It seems, George, that your physician prescribes plenty of fresh air and sunshine for you, during the next few days. Do you think you are strong enough to-day, for another mobile excursion over the farm?"

"Yes Fillmore, quite strong enough, provided the excursion is not too long. To-morrow, if the weather should be fine, I hope we may be able to take that trip to Fenwick, which you spoke of on the afternoon of my arrival. The more I see of the farm, the more I am interested and delighted. In a very short time, I believe I might become an enthusiast on the agricultural question. Hitherto, I have had an unexpressed antipathy, towards farm work.

"Strongly impressed with the idea, that a farm life must necessarily, be as dull as ditch water; I find Solaris a revelation, which has opened my eyes and scattered my foolish prejudices to the four winds. At every turn, some new surprise awaits me. My typical farmer, with his shock of untrimmed hair and beard, his stooping shoulders, his shambling, plow-following gait, his great cow-hide boots, his coarse, soiled, slouchy, ill-fitting blouse and overalls, his grimy hands, his ill-at-ease, uncultured manners, and his born-tired expression of countenance, I cannot find. In his place, much to my astonishment, I do find a splendid people, in the prime of life, lithe, active and energetic, in the possession of a superabundance of vitality, which gives them the graceful air of having grown to a perfect maturity, on the sunny side of life. What does it mean? Everywhere, I am politely greeted, by dignified, graceful, self-poised, rosy-cheeked, bright-eyed, happy, well-dressed, educated, refined and polished men and women. Can it be possible, that they are farm laborers?"

"Every one, friend Gaylord! It is to rightly organized farm labor, properly supplemented by appropriate machinery, that these people owe the superior condition in which you find them."

"You have surely created a new era in farming, Fillmore! Do you think a general introduction of co-operative farming, will produce equally successful results elsewhere?"

"Much better and more satisfactory, George! Co-operative farming, even here at Solaris, has as yet scarcely passed the threshold of the experimental stage. Every new farm, will profit by the errors and successes of those previously established. Each one will add to the strength and working capacity of the mass. This improvement will steadily increase, until the children born under the new system, become its principal working factors. When that time arrives, the influence of the born and bred agriculturalists, will have grown so strong, socially and politically, that a new impetus will be given to the movement, by the favorable legislation which they can then command.

"When we consider the future of the co-operative farm, as a working factor for good, in the affairs of the Republic; we can then appreciate the great importance of the movement. Stirpiculture, wedded to agriculture, ushers in a new era for the birth and education of an epoch-making race of dominant thinkers, so well born, so self-poised, so harmoniously developed, physically, intellectually, and spiritually, that without effort, they are naturally chosen by the masses, as social and political leaders."

"What an enthusiastic dreamer you are, Fillmore! The picture of the future of the movement, which you have so graphically drawn, seems too good to be true! My brain is in a whirl trying to follow you! Let us now prepare for that promised ride."

CHAPTER XL.

THE COMING ERA OF GOOD ROADS.

"SINCE our mobile excursion to the farm village of Fenwick, I have been haunted by the beauty, smoothness, utility and durability, of the magnificent highway, which now connects the two villages. I am more than ever impressed with the power of the co-operative movement, to effect a revolution in all industrial methods; especially, in travel and the transportation of farm products. Tell me, Fillmore! Do you think this road-building fever, will continue to spread with the growth of the movement?"

"Yes, George, with every new road, will come an added impetus to the movement, which will insure a steady progress. The importance of good roads as a source of wealth, and a mark of civilization, is just beginning to be understood by agricultural people, and by rural populations generally. Oppressed on every hand by the universal extortion of railroad monopoly, they are slowly awakening to a realization of the fact, that the question of cheap transportation, is for them, the one, overshadowing question, which demands immediate attention.

"As an object lesson on the subject of good roads, the introduction, and constantly increasing use, of bicycles, motor cycles, motor freight wagons, automobiles, electro mobiles, locomobiles, and the entire class of vehicles equipped with rubber tires, has aroused a widespread interest, which is prophetic of great results. Acting as a strong reinforcement to this educational work, the co-operative farm, with the advantage of its village organization, representing in the public mind, such an attractive combination of agricultural, industrial and social life; will by the force of example, give an additional impetus to the systematic construction of broad, permanent highways; that shall prove a source of pride, to the community through which they pass; roads, that shall last for centuries.

"Reacting favorably, in broadening the mission of the co-operative farm-village, with its promise of permanent homes, and employment for the unemployed, and the homeless; the continuous construction of these free avenues of travel and transportation, will soon affect the status of all rural populations, by vastly increasing their wealth and power. For them, the vexed problem of transportation, will be solved. They will discover by actual experience, that these wide, durable wagon roads, will connect them with distant centers of traffic, and serve them better and more honestly, than steam railroads; that in cost of construction and repair, they are much cheaper; that when constructed, they belong to the people as absolutely, free highways; that no greedy corporation, can control them; that no threatening, irritating, lawless force, of Pinkerton's armed thugs, is required to protect them; and finally, that they offer every inducement to unfettered genius, to invent and to freely exploit, better and cheaper vehicles.

"As one grand result of this combined educational work, rural life will become exceedingly desirable and charming. The great city, will lose its attractive force. The tide of migration, will flow back to the pure air, invigorating sunshine, blue sky, and the verdure-clad hills of the country. In a general way, we may predict, that a few years hence, everywhere throughout this broad land, we shall find picturesque, prosperous, well populated villages. As the minor centers of education, art-culture, refinement, amusement, progressive race-culture, scientific agriculture, esthetic, social and co-operative life; they will be embroidered, like a vast net-work of shining pearls, on a perfect system of broad, smooth, highways In their construction, ornamentation and maintenance, these good roads will utilize and express, the pride, energy and best inventive genius, of the village centers thus linked together. As a result, the Republic will be gridironed with a superb system of free highways, more permanent, more perfect, and more beautiful, than those old, historic, Roman roads, which even now are existing monuments to the solid character of Roman civilization.

"This imperial road system will be complete, when the co-operative farm has reached every township in the union. Then, we may calculate the results, which are to follow. Broad, tree-shaded, park-lined, flower-bordered boulevards, will connect New York with San Francisco; Galveston with Saint Paul; Portland, Maine, with Portland, Oregon; Los Angeles with Saint Louis; Boston with Buffalo, Philadelphia, and Baltimore with Jacksonville, Florida; New Orleans with Cincinnati and Chicago; the wonders of Yellowstone Park, with the crags and glens of the White Mountains, Niagara Falls, with the Grand Canon of the Colorado; the orange groves of Florida and California, with the picturesque, cool, invigorating, health resorts of Lake Superior; the wheat fields of the great Northwest, with the coal mines of Pennsylvania; Washington, the nation's capital, with every seaside resort, every mountain view, every beautiful city, every healing spring, and every hamlet and village of the Republic.

"Pulsing with a new tide of social and industrial life, flowing through the arteries of this unequaled system of great highways ; all of these places, both great and small, will become more closely bound together, by the links of a new social order ; representing the beginning of a higher civilization. Then, these beautiful highways, will be glorified and appreciated by mankind, as the monumental work of one, broad system, of co-operative farm villages. Then, these villages, which have made such a system possible, may collectively claim the proud distinction, of being known as the Nation's Committee on Good Roads."

"Excellent! Most excellent! Fillmore. Your prophetic vision, with the vastness and the brilliancy of its sweeping scope, fairly takes my breath! Yet, I must confess, that judging from the masterly system of road-building inaugurated by Solaris and Fenwick, the evolutionary results which you so confidently predict, are both reasonable and logical. What additional results, do you claim for the system?"

"At this time, George, neither tongue nor pen, may attempt to describe the marvelous results which will follow the introduction of an era of good roads. In a brief way, I will try to give a few of the most important. In the matter of travel and transportation, these free highways, will annually, save millions of dollars to citizens of the Republic, by enabling them to escape from the clutches of the largest and most powerful of all monopolies; the railway monopoly. A monopoly, that for many years, has held the public by the throat; exacting a tariff so exorbitant, as to be almost prohibitory. A monopoly, which has had the amazing gall to pose as the farmer's especial benefactor. A monopoly, that while so posing, has robbed the country of one-half its wealth, by transferring the same to cities. A monopoly, that in the name of good business, has had the stupidity to decree through its tariff schedule, that miles and miles of empty freight cars, shall daily, throughout the land, roll past hundreds of thousands of farms, where countless tons of heavy freight, in the way of fresh vegetables, lie rotting for the want of a market. A monopoly, that never neglects an opportunity for fleecing the public. A monopoly, so unscrupulous, that for the pork trust, it will haul a hog across the continent for ninety cents; while for indifferent service, it dares to charge the people, from two and one-half, to five cents per mile.

"And yet, George, just think of it! In the beginning, this monopoly was chartered to serve the people who granted the franchise. A monopoly, now grown so bold, that when the public protests that the franchise is violated, because the interests of the people are no longer served; a Vanderbilt railroad king, insolently replies: 'The public be damned!' A monopoly that has killed all healthy competition, by organizing all railroads into one giant pool; thereby creating the mother of trusts, controlling a corruption fund of enormous magnitude. A monopolistic trust, grown so rich and powerful, as to be beyond the reach of law; boldly corrupting courts, buying legislators, and turning the administration of justice into a farce. In fact, this monstrous combine, has become so dangerous to every interest of good government, that the law of self-preservation demands that it shall be speedly wiped out, by the government ownership of all railroads.

"We may now consider the ways and means, by which our co-operative system of good roads, can control railroad freights, and finally drive railroads to government ownership. Long before the close of the first half of the twentieth century, thousands of miles of these fine wagon roads, will be found in every State. Responding to the demands of legions of voters, who reside in the co-operative farm villages bordering these charming highways ; a strong force of legislators, will everywhere rise up, as eloquent advocates of the good roads movement. Honest and faithful, inspired by a tenacity of purpose which will brook no opposition from railroad lobbies; encouraged and strengthened, by an ever increasing army of enthusiastic voters behind them, these tireless legislators will not halt, until the entire system of good roads, so well begun by the farm villages, shall be taken up, completed, and perfected by the State. Ten years of such forceful work, will surely accomplish the task.

"Then, to the champions of the system, shall come their reward. They shall behold, flowing in mighty streams, over the wide, petroleum treated, dustless surfaces, of these far-reaching, absolutely free highways, the traffic and travel of a mighty Republic!

"Then, will come the demonstration of what American genius can do, toward the evolution of a superior class of rubber tired, horseless vehicles, which shall prove the best, cheapest and most durable, for purposes of freight, traffic, and travel, on such a complete system of fine roads. The best of our present types, when compared with these twentieth century road flyers and freight rollers, will seem poor, crude affairs. The irresistible volume of this swift stream of the new travel, and the new transportation, eloquent with the progress of the century, will herald the coming of a well-merited doom for the monopolistic railroad combines.

"Then, local travel and traffic, will make haste to desert the iron rails. Railroad freights everywhere, will fall to zero. Short railroads—branches and feeders to main lines—will become useless and worthless. Many of them will be sold at auction, for less than the cost of the iron in the road-bed.

"Then, shorn of their ill-gotten gains, the mighty railroad rings of the land, will fall from their tall pedestals of pride, where for years, they have posed as owners of the earth. With financial ruin staring them in the face, they, and the whole brood of erstwhile railroad kings, will make urgent haste to sell to the government, at the bare cost of construction, such great through lines as may be necessary to maintain inter-state commerce, and across-the-continent traffic. Other roads, they may not sell at any price. A government for the people, and by the people, will have no further use for them.

"Then at last, the supreme folly of having a half-dozen competing lines, running side by side through the same territory, will be fully demonstrated. With this demonstration, will come the opportunity, to scores of paid press writers, pessimistic bigots, self-conceited, unprogressive wiseacres, who have so long and so loudly derided the government ownership of railroads, as the most suicidal and unbusiness like scheme ever hatched; to answer this pertinent question: Would it be possible, for government engineers building public railroads, to ever be guilty of such monumental stupidity?

"The social effect of these good roads, on the

lives of all agricultural people, will prove even more important than the financial advantages gained. Hitherto, they have been so hampered by environments, by lack of means, and lack of leisure, that as a class they have been unable to enjoy or to appreciate the wonderful, the educational, the broadening and the refining effect of much travel, on the mind of the individual. From lack of experience, they do not realize that the sum of human life is the sum of its sensations, which are produced by change of environment, contact with a larger or lesser series of natural phenomena, and more especially with other lives.

"The more progressive lessons of life, are learned from example and not from precept. Men and women, are only children of a larger growth, they are imitative creatures with a natural instinct to choose other, higher, and better lives as models. Hence the great value of travel as an educator. The larger the area covered by the traveler, the wider the field of experience and choice. Through the law of action and reaction, social contact with a multitude of actors and thinkers, refines the individual. A healthy spirit of emulation is aroused, which leads on to progress.

"With the advent of a universal system of good roads, cheap travel, and a dominant combination of co-operative, industrial and agricultural enterprise, an extraordinary era of recreation and travel, will dawn for all rural people. Opportunity, leisure, and means will be abundant. All co-operative workers, can afford to take an annual vacation of at least one month. The ownership of a swift, roomy, durable, road machine, capable of making from twenty to thirty-five miles an hour, will be within the means of every family. In this private car, the family, or a select party, could easily and leisurely accomplish a five thousand mile tour in twenty days. Along the whole distance, farm villages, from fifteen to twenty minutes apart, would offer the travelers, machine supplies, repairs, and excellent hotel accommodations, for an expense not in excess of the same at home. Than this, no traveling excursion could be more delightful! For pure enjoyment, a select party of nineteenth century millionaires, could not equal it.

"The enjoyment of such delightful opportunities for even a single decade, would make the rank and file of the republic thoroughly acquainted, with the soil, scenery, forests, lakes and rivers; the mining and manufacturing possibilities; the peculiar characteristics of the people, their local ambitions, political wants and future demands, of every state and county in the union.

"Thus equipped with this important knowledge, each voter, both men and women alike, would be prepared at any time to vote intelligently and wisely, on every question affecting the welfare of the republic as a whole, or in part. Elected to Congress, these voters would appear as the ablest, most patriotic, most just, and most incorruptible body of law-makers ever known. Understanding the equities of righteous dealing between themselves as fellow citizens, they would be prepared to decide correctly on all questions of an international character, which might affect the interests of the world at large. This would be a demonstration of the rule, as to the formation of a true republic. To make the entire political fabric both enduring and progressive, the units or voters, must be well born and rightly trained. Of this training, travel is an essential part, which should not, which must not be overlooked.

"As affecting their social and intellectual progress, these years of travel would improve all classes of agricultural and industrial people, to a still higher degree than the one achieved in political expression. A general interest would be aroused in questions of political economy, race culture, psychology, and physiology; geology, geography and history, botany, chemistry, and mineralogy; which later, would lead to close reading and hard study in the whole domain of scientific research, as the one sure method of increasing the scope of individual happiness. Every succeeding year of this travel-training, would result in binding all classes still more firmly together, into one harmonious, homogeneous mass. Now George, tell me what you think of the good-roads question! Is it not one affecting the vital interests of humanity to a marvelous extent?"

"Marvelous, Fillmore! Most marvelous! Hereafter, you can count on me as an enthusiastic advocate. I cannot say too much in its favor."

CHAPTER XLI.

CO-OPERATIVE ETHICS.

"SPEAKING of wages," said George Gaylord, "did I understand you to say, that all of the co-operators at Solaris receive the same pay?"

"Yes, George, equal wages for all classes of workers, is the motto at Solaris. Recognizing the solidarity of the interests of society, simple justice demands the same rate of pay for each member of the company; without regard to sex, or particular qualification."

"It seems to me, Fillmore, that justice would demand that each one should be paid according to skill and capacity. I cannot understand, how anyone capable of being a foreman, would be content to accept, as a just equivalent for his services, a compensation as low as that awarded to the least capable worker in the colony."

"I think I shall be able to convince you, George, that a correct view of this question, is largely a matter of education. You have, perhaps unconsciously, voiced the usual argument against the equity of equality, which is made by the champions of the competitive system. Our people have learned from experience, that the co-operative farm movement is a leveling up process, which purposes to raise the weaker units, to the condition of the higher. They have learned, that society is a purely co-operative institution. They have learned, that the wants of society, create value for the products of labor. Society, then, is labor's market. In this market, the wants of the weaker units, are just as important, as are those of the stronger. Stimulated by the number and variety of these wants, inventive genius has given to us tools and machinery, which have increased, at least one hundred fold, the capacity of labor to produce. In the creation of tools and machinery, the mental acuteness and inventive skill of the weaker unit, often surpasses that of the stronger. It follows, then, that each one of the weaker units, is justly entitled to an equal share of the advantages which are conferred on labor by society, with its market and equipment of tools and machinery. These advantages, make the productive work of all classes, nearly equal. Let us try to find the real difference, between the daily labor products of the strongest and the weakest workers. Let us consider present conditions here at Solaris, as an illustration. Let us take one hundred dollars, as the value of the product of one day's labor, by an average person, plus the advantage of such superior social organization, training, tools and equipment, as Solaris can now furnish. On the other hand, let us take fifty cents, as the value of one day's labor, by the strongest, most capable worker, when isolated from his fellows, and from all social organization, with its tools and equipment. Under the circumstances, allowing that the strongest could produce twice as much as the weakest, we should have twenty-five cents, as the value of the daily product of the weakest worker. These sums, compared with one hundred dollars, would give us the exact difference between the strongest and the weakest, under the favorable co-operative conditions, existing at Solaris. A difference, so trifling as to be scarcely worthy of consideration, only one-fourth of one per cent. What think you, George! Where now is the injustice of equal wages? Remember, when justice is done, the mission of charity is finished!"

"Your clear statement of the case, has proved a revelation to me, Fillmore! I am quite ready to acknowledge the exact justice, of your co-operative system of equal wages. I am profoundly impressed with the soundness of your argument, that women and all weaker units in the army of labor, are justly entitled to an equal share of the advantages conferred on labor, by social organization, and by the education, training and equipment, resulting from that organization. This view of the question, is a new one to me. It places the whole subject, in quite a different light. By the aid of this light, I am beginning to understand something of the intricacy and force, of this co-operative machine, which we call society; and how much it affects the question of labor and wages.

"My experience with co-operative farming here at Solaris, is beginning to bear fruit. Under your instruction, friend Flagg, I think I can now understand the wide difference, between the competitive and the co-operative systems of organized labor. The former, benefits the few at the expense of the many. The latter, raises the individual, by benefiting the mass. The first, seems to be a constant menace, which threatens the peace, welfare and stability of society; clearly making for evil. The second, striving for the interests of all, builds up, strengthens and purifies the weaker units; unmistakably making for good. The results seem to marshal themselves on the side of co-operation, for the purpose of demonstrating the truth of its shibboleth, that the injury or weakness of one, is the concern of all. In other words, to raise the lower strata of society, means a corresponding elevation for the upper. The average morality, happiness and prosperity of society, is measured by the morality, happiness and prosperity of its weaker units. Tell me, Fillmore, does the acceptance and advocacy of this view of the relations existing between labor and society, make one a socialist?"

"They surely do, George! They make you a socialist of the most progressive type. I am both surprised and delighted, to find how well you have learned the lesson of co-operation."

"If the co-operators at Solaris, are socialists, then

they must be good people. I am perfectly willing to be classed with them. At all events, I am a thorough convert to the co-operative system. I can now understand the scope and significance of the work; and why it is, that the Solaris workers, are so much superior to any farm people I have ever known. I begin to perceive that the success of the co-operative farm, means the regeneration of society.

"This morning, Fillmore, under the guidance of Miss Houghton, I visited the kindergarten, the schools, the club rooms and the theatre. I was amazed, to find such a magnificent system of education and amusement, in successful operation, for the benefit of a farm village. Indeed! A city of fifty thousand people, would be very fortunate, in the possession of such a fine one! How did you manage to make it possible?"

"In carrying out the wise plans of Fennimore Fenwick, you behold to-day, the result of combined co-operative agriculture and stirpiculture, which affords to our people, and to their children, conditions for education and amusement, fully equal to anything, money can procure for the wealthy. Children born at Solaris, under carefully prepared conditions for a perfect motherhood, are endowed with a precious birth-right, far superior to anything heretofore known to heirs of wealth. The system is being constantly improved. As it now stands, I consider it the crowning success of the co-operative movement.

"Speaking of Miss Houghton, George, reminds me of a question! You have yet to tell me, the result of your first interview with her. Did she seem to blame you so very much, for not answering her mother's letter?"

"Oh! no! She was kindness personified. She hastened to assure me that, in the light of subsequent events, she came to understand the whole situation. It appears, that after writing the letter in question, her mother grew very much better. In this improved state, she lingered for some time, and did not die until several weeks after Miss Houghton had read to her, the notice of my mother's death, which came to them through the columns of an occasional New England newspaper.

"Having answered your question, Fillmore, I will now return to the subject of my visit to the schools. The interest manifested by both children and teachers is something to be proud of. The amount of general information of a practical character, which the pupils have acquired, even in the lower classes, is quite surprising. This is especially noticeable, in the ready knowledge they display, regarding current political events; including the personal history, character and ability, of the various political leaders. Is it wise, to devote so much time to teaching politics; and to commence this teaching with children so young? Do you really consider it so very important?"

"Yes, George, it is a matter of the utmost importance! A republic of ignorant people, is a republic only in name; in reality, it is an oligarchy. On the contrary, a true republic, is one in which all its units or voters, are so educated, that they are familiar with the theory and practice of government. They must know that true government is a co-operative institution, which must guard and protect with exact justice, the interests of all of the governed. They must know, the extent and condition of the agricultural, manufacturing, commercial, mineral and lumbering resources of the country. They should understand diplomatic, domestic and foreign relations. They should know every detail, of the educational, financial and political wants of the masses, in the domain of each State or Territory. Finally, they must be familiar with the character, trustworthiness and ability, of all political leaders. Children of the co-operative farm, are educated and trained, in a manner that will best fit them to become true citizens of such a republic. This is why, a practical, political education, to be successful, must become a matter of interest to the children while they are young. They will then learn, that a true republic, is a co-operative machine, which cannot run smoothly, while one imperfect cog remains to retard the action of its wheels. This valuable lesson, they cannot learn too soon. What think you, friend Gaylord?"

"I cannot quite agree with you in this matter, Fillmore! I think it would be far wiser, while they are so young, to teach these children such lessons as will give them the ground work for a sound religious faith. Then they will understand the first importance, of being prepared to save their own souls. Later, in the closing school years, they could be taught your progressive, political scheme, which I think is a remarkably good one."

"Stop one moment, George! I see Miss Houghton is coming. She will be delighted with an opportunity to answer some of your objections, to the co-operative code of ethics, evolved by the people of Solaris."

"You are a welcome visitor, Miss Houghton! You have arrived, just in the nick of time! Our mutual friend here, Mr. Gaylord, has been telling me of his visit to our schools, under your guidance. While he praises the wonderful progress made by the pupils; he seems to think, that we teach too much politics and too little religion."

"Pardon me, Miss Houghton!" said George Gaylord, "I assure you, that I was not indulging the spirit of fault finding! Allow me to explain! I had reached a point in our discussion, where I was about to remark, that since Adam's time, the people of the world have been born, heirs to the dominancy of total depravity. With this heritage, we are as prone to sin, as are the sparks to fly upward. Under such circumstances, it would surely be the height of folly, to attempt to overcome this natural tendency toward evil, without the aid of the strong arm of the church, with its broad mantle of christian faith and saving grace."

"I grant you, Mr. Gaylord, that with your peculiar training, such a conclusion would be quite natural."

"Now, Mr. Flagg! I have a word for you! We must make every allowance, for Mr. Gaylord's theological education. An education, that has filled his mind with somewhat distorted meanings, for the terms, religious faith, soul, sin, salvation, religion, total depravity and many others of a similar import, which theology has applied to man's spiritual welfare. Just at present, the difference between us, is wholly a matter of definition. When we have acquired a true meaning for these disputed terms, we shall stand harmoniously on a common ground. We shall then be ready to accept the higher teachings of the new religion. A religion of spiritual evolution and unfoldment, which responds to the progress of the twentieth century."

"You are quite right, Miss Houghton! I am very

willing to make the generous allowance you suggest. I think Mr. Gaylord would be glad to hear your views, regarding the practical teachings of the new religion."

"Thank you, Fillmore!" said George Gaylord, "you have voiced a request, I was about to make. I trust Miss Houghton, will proceed at once. I will promise to be a listener, who is both interested and attentive."

"I will promise one thing, Mr. Gaylord. It is this, before I have finished, I shall do my best, to convince you, that in embracing the new religion, the people of Solaris have devoted themselves to a system of religious teaching, which is far too broad for the limitation of church walls. That this new religion, is so practical, and so exacting, that its followers, if they are true, are in duty bound to observe it as a rule of life, seven days in the week, year in and year out.

"As a primary basis, the new religion teaches, that all human life is sacred. That it is the highest expression on this planet, of an Omniscient purpose. Conscious life, or the capacity to become conscious of anything, is a Deific attribute. All knowledge comes to the mind through the avenue of the senses, or from sensations produced by contact with existing things in the domain of Nature. The domain of Nature, is the domain of the Omniscient! All real knowledge, acquired from this domain by right methods, which is in harmony with natural evolution, is Truth. Truth, then, is Divine!

"From these broad premises, we may deduce, that to acquire knowledge, or to accumulate truth, becomes the highest duty of life, a religious activity of the highest order. To be engaged in the intellectual process of gaining knowledge, is to be engaged in a spiritual work. The intellectual process, is a spiritual process. By the psychologic action of the mind, through its sub-conscious functioning, all knowledge coming through the senses, first becomes the spiritual possession of the Ego, the Soul, the seat of consciousness, before it can be expressed materially by the mortal man. Hence, spiritual evolution, is a natural growth, a crowning part of physical and intellectual evolution. The body, as an associated colony of more or less intelligent cells, is an important part of the thinking machine. Body, brain and intellect, in their dual existence on the material plane, form an important trinity, which enables the Spirit to accumulate knowledge, and also to retain that knowledge, after the passing of the physical. To dispute this postulate, would be manifestly absurd, as the spiritual man is the conscious Ego, the real gleaner and possessor of knowledge. It follows then, that to be engaged in any kind of educational work, is to be engaged in a religious work of great spiritual importance. That, through proper intellectual training, we may obtain spiritual growth, rebuild the moral character, exterminate vice, and unfold the graces of virtue, purity, honesty and goodness. These are spiritual attributes, which embrace all there is in the domain of morals.

"In appealing to the new religion, for a broader, truer definition of the term, Soul, we learn that Soul, as a cosmic unit of the larger cosmos, is the repository of infinite possibilities: That evolution is the law, by which these possibilities are unfolded: That it inherits immortality as a birthright, from the Great Over Soul, the source and center of all life: That, in fulfilling the law of life, by sojourning in the flesh for a brief period, it cannot be lost, or become totally depraved; although the body, which is but its earthly expression, may become so debased by poverty, selfishness and sin, as to momentarily thwart the Divine purpose of life.

"From the same source, and by the same authority, in response to a sincere desire for a better definition of the word Sin; we are taught, that the object and purpose of the existence of this planet, is the evolution and perfection of the human race. Human life, then, is the flower and fruit of the planet. As such, it is the direct expression of a Divine purpose. At the command of a higher law, this life must at all times, be treated as sacred. From this high rock of observation, we perceive that all acts, by society or individuals, which tend to promote, protect and purify this life, are helpful along lines of evolution; therefore, righteous and good. In their doing, these acts become the highest expression of a religious duty. On the contrary, all acts, by society or individuals, which tend to destroy, injure, poison or sully this sacred life, or to bar its ordained progress, are in themselves, unholy, wrong and criminal. In commission, these acts become the greatest of all sins. The logic of this deduction, is beyond dispute; because they are direct attempts to thwart the progressive and evolutionary purpose of the planet; therefore, they must be considered as sins of the first magnitude.

"Second in magnitude, and akin to these in wickedness, is the sin of society against women. A sin so potent for evil, that at the behest of selfishness, greed and lust, government, church and society, with one accord and without a protest, join in denying to woman an existence of financial independence. This denial makes slaves of women, who should be noble, pure, self-poised, self-sustaining and absolutely free. But the acme of wickedness is reached, when this denial reduces women to creatures of merchandise, when every year, it drives unnumbered thousands of them to lives of degredation and shame; thus perpetrating the crime of the century against unborn generations, by tainting and poisoning the fountain of life at its very source. The new religion has decreed, that the mothers of a perfected republic, must of a necessity, be both pure and free. It purposes to cure this crime, by working through the strong arms of an ever-increasing series, of unselfish co-operative brotherhoods, where a progressive union of agriculture, and stirpiculture, shall provide for and protect both mothers and children; at the same time furnishing the ways and means, which offer an honorable, useful self-sustaining existence to all woman kind, be they wives, mothers, sisters or sweethearts.

"Third in magnitude and closely allied to the first two, is the great sin of ignorance. The mother of bigotry and superstitious fear; the father of duplicity and craven cowardice! What we know, we fear not. It is only the mysterious darkness of the unknown, that is filled with terror. To abolish ignorance, is to make the mind master over matter. Mind is both the spiritual and the intellectual expression of the soul. True culture of the mind, is moral culture. It is only the well grown, highly cultured mind, that can reflect the inherent graces of the spirit, which mark all noble characters. To the individual, who has acquired a knowledge of the law of evolution and environment, is given the power to control environmental conditions; by wresting from nature the secrets of success, in feeding, clothing, housing, educating and elevating humanity. It follows then, that to overcome the sin of ignorance, is to banish poverty. To banish poverty, is to banish want. To banish want, is to take away the very foundations of the sin of selfishness. Selfishness, is the father of a multitude of sins, which must perish with it.

"From these premises we must deduce, that all educative work in the proper sense, is a religious activity, which makes us better acquainted with the relations which exist, between man and his Creator, the Great Over Soul. The spiritualizing influence of this intellectual work, carries with it the compensation of a great reward. It crowns the gleaner, with happiness of the purest type. As knowledge increases, the field of knowledge expands, the flood of happiness swells in volume. A long busy life on the material plane of existence, is far too short to acquire this vast treasure, which is commensurate with the needs of progress for an eternity of spiritual existence, to which, this life is simply the primary school. With a better understanding of the nature of sin, and of the alarming extent of its evil influence over human life; the new religion undertakes to bless mankind, by banishing ignorance, poverty and crime. To this practical, spiritual work, the people of Solaris religiously devote themselves, as being a life-work of the noblest order.

"The three principal sins which we have considered, may be justly regarded as the parents of all lesser sins. Having given a few brief suggestions as to methods of cure, which are offered by the new religion; I am now ready, Mr. Gaylord, to take up the doctrine of total depravity; which plays such an important part in your theology.

"As the primary step, I will re-state a prior postulate, as follows: The spiritual man, is the conscious Ego, the Soul, or a cosmic unit of the larger cosmos; an indestructible part of the great life principle. As such, it is the repository of infinite possibilities, which are destined to be unfolded by the law of progressive evolution. From the Great Over Soul, it inherits immortality and indestructibility; therefore, it cannot be lost, saved, or become depraved. The mortal body is an outer covering, through which it must express itself on the material plane of existence. Physical, intellectual and spiritual life, are subject to the law of evolution, by which they achieve progression and fulfill the purpose of existence.

"To assume, that the people of this planet, are born subject to the dominancy of total depravity, is to deny immortality, and the truth of these postulates. In denying them, it denies the existence of a dominant principle of good, and affirms the existence of a dominant principle of evil. It also denies all progress, all moral reform, every noble aspiration, every good deed, all evolution, all science and all reason. Where then, in the economy of nature, is there room or use for the doctrine of total depravity? A doctrine so pernicious, that in the mouths of its advocates, it has done more than aught else, to destroy the confidence of mortals, in the wisdom and justice of the Divine plan of the universe. To even assert its existence, is to question the existence of a universe, under the reign of justice, law and order. Evidently, the doctrine of total depravity, does not belong to the domain of fact. It is equally clear, that it must be a theological fiction. A sin of theology against progress, which in the dazzling whiteness of the spiritual light of the new religion, must soon fade into oblivion.

"Can we teach politics to school children, as a part of our religious duties? Is a question we will now consider. The answer, will depend largely on the definition, which we give to the word religion. Let us try to find a true definition, broad enough to embrace an affirmative answer to our question. As a basis, we have human life as the highest expression of the planet. With the physical body, as the basis for intellectual evolution. With intellectual evolution, as the basis for spiritual evolution. Hence, we have as a conclusion, that the spiritual development and unfoldment of the race, up to a point where it can accept the truth of immortality, is the logical purpose to be accomplished by all religions. Reasoning from these premises, it would seem clear, that the practical value of any religion, must be measured by its ability to teach the people how to help themselves; how to master the great problem of physical life, by attaining perfection in the arts of feeding, clothing, housing, educating and spiritualizing the race. If, in connection with these solid foundations for a natural religion, we add the important fact, that this is a republic, in which the wish of the majority, should become the law of the mass; we shall discover that politics become the natural channel, through which the wishes of the majority are expressed; that corrupt politics, result in bad government; that pure politics, insure good government; that a wise, just government, is the greatest political benefit which can be conferred on the people governed. United, these conclusions give an affirmative answer to our question. They also tell us why, the new religion, the mouth-piece of inspiration, reason, science, evolution and progress, should proclaim it a religious duty, to teach our children, —embryo citizens of the republic—every practical detail of pure politics.

"What think you, Mr. Gaylord? Have your objections, been satisfactorily answered? Can we agree to accept new definitions, for the disputed religious terms, which we have been discussing?"

"I am satisfied, Miss Houghton, that I have been quite too hasty in my conclusions! You have convinced me of the importance of teaching pure politics to children, as a part of their religious training. With regard to other religious questions, you have answered my objections in a most masterly manner! The practical religion, which you have so beautifully outlined and so clearly defined, seems worthy of all the eloquence which you have bestowed upon it. That dreadful doctrine of total depravity, which you have so effectually demolished, has always been a repulsive one to me! For years, it has been a tormenting theological thorn in my side! I could never quite reconcile its existence, with the overruling dominion of an all-wise Creator; the very embodiment of Infinite goodness. I may as well say frankly, that I have often tried to find some good reason for denying it! Now, I have found one, that will satisfy my conscience. With the vexing doctrine of total depravity eliminated from the religious problem, a definition for the term, practical religion, becomes much more simple. A new light is thrown on the whole subject. Just at present, under the influence of this light, I am inclined to think, that your statements and your premises, are all true. Granting this, I will cheerfully admit, that the people of Solaris, are nobly living practical religious lives.

I am very much interested in the wonderful claims of this new religion. I trust, that after some weeks of careful examination, I may be able to accept them without one single reservation. After that, I venture to promise, that we shall be able to agree on a satisfactory definition, for all disputed religious terms."

"Bravo! George! Now, you are talking more like your old self, more like a reasonable man. You are making great progress, in mastering the underlying principles and practical work of the co-operative movement! I think, Miss Houghton, that you ought to join in offering congratulations. Will you not?"

"Yes, Mr. Flagg! I shall be glad to do so! First, I want to compliment Mr. Gaylord, on his excellence as a listener! Then again, I wish to thank him, for his kindly summing up, of the impressions, which came to him from my rather long sermon on practical religion.

"Now gentlemen, you must excuse me! I have an engagement, which demands my immediate presence at the kindergarten."

CHAPTER XLII.

RURAL LIFE UNDER THE REIGN OF CO-OPERATION.

"I WISH, Fillmore," said George Gaylord, "to question your statement, as to the ability of the co-operative movement, to check the rush from country to city life. The tide of the movement is a strong one, that has been constantly increasing in volume, for the past twenty years. I fear that even the popular co-operative movement, will fail to turn the flood."

"The thing is sure to be accomplished, George! But, to understand the workings of the underlying force, which shall make this change possible, we must first study the units of rural society. Of course, the financial basis of these units, must be supported by agriculture. Agriculture is, and must continue to be the main support of all rural populations. Fifty years ago, agriculture as a whole, comprised a vast collection of small farms and farmers. Then, the small farmer and his family, as the stable unit of suburban society, was financially and practically independent. Questions of over-production of food products, rise or fall in the price of exchange, panic in the money market, or an adverse balance of trade, disturbed them not.

"Under the spur of necessity, and as a part of the legitimate farm work, the farmer and his family, in a crude way, practiced many of the industrial arts, such as leather working, harness making, boot and shoe making, cloth making, the carding, spinning and weaving of wool; the preparation, spinning and weaving of flax or linen fabrics; the manufacture of many farm implements, brooms, baskets, harrows, sleds and carts; tailoring, making all kinds of underwear, hosiery, gloves and mittens; linen furnishings, for table and bed, together with many other articles of household use. Often, the forge and the anvil, with tools for rough iron working, were added to the equipment of the farm. In those days, farming required a knowledge of the use of tools; the square, the level, the plumb-bob; the hammer, the saw and the plane; were as necessary to the farmer, as they were to the carpenter.

"If we carefully study the significance of these things, we shall soon discover, that in reality those farms were practically, combined agricultural and manufacturing institutions, which were self-supporting and self-sustaining to such an extent, that farm people were the most independent on the face of the globe. As such, these small farm centers were potent factors, in swiftly advancing the permanent wealth and civilization of rural society. Born and trained in this practical school of life; financially unshackeled, therefore politically free; our farmers of fifty years ago, developed a spirit of sturdy independence, a patriotic devotion, a steadfastness of purpose, a self-confidence, and a power of the initiative, which made them the pride and the bulwark of the nation. They were the well trained, trustworthy citizens, of a true republic.

"Evolutionary progress, moves forward by waves. The depression between the crest of the last and the summit of the succeeding wave, represents the transition, from one step of progress to the next higher. Therefore, periods of depression, need not cause alarm, they are in reality prophecies of progress. Let us apply this evolutionary law to agriculture and its people, as being in the transition stage, during the past forty years.

"Since the beginning of the last half of the nineteenth century, the separation between agriculture and manufacture has been going forward, the gulf between them becoming wider and more absolute, with each succeeding year. Invention, improved machinery, combinations of capital, the sub-division of the various trades into specialties, leaving the worker, master of none; all have served to develop the entire system of manufacturing industries, to a degree out of all harmony with the tardy progress made by agriculture. The mining and manufacturing craze, has swallowed up all other interests. Like a whirlwind, it has spread over the land, drawing into the ranks of its toilers hosts of agricultural workers; thus swelling the army, producing manufactured articles, and correspondingly reducing the home market for such things.

"These conditions have naturally produced a congested market. Logically, there has followed, periods of stagnation, labor riots on account of reduced wages, periods of enforced idleness, and panics in the money market; all culminating in a loud demand for relief from the burden of over-production, by securing control of foreign markets. So completely has the manufacturing craze dominated the commercial and political economy of the republic, that both leaders and people are blind to the real cause of the calamity. An aggressive and progressive minority begin to realize, that the laborer and the farmer are no longer free, that they are the slaves of capital with its factories and machines, or of railroad combines, which control all lines of transportation. But no one sufficiently understands the situation, to be able to answer why.

"Now let us study the history of agriculture, during the past forty years. This trying period of transition, has been marked by many changes. The small farm family, shorn of its ability to manufacture, even in a crude way; for shoes, clothing, bedding and table linen, must patronize factories located in distant cities. In order to pay for these things, much farm produce must be shipped to remote markets. In both cases, such heavy freights, commissions and profits, are paid to lines of transportation, middle men and handlers, that at the end of the year, the farmer's net proceeds are reduced to zero, or at least very close to that point. If the farmer be in debt, he finds himself unable to pay the interest on the indebtedness. If the farm represents much invested capital, the net income of the farm becomes too meagre to pay even a moderate rate of interest on its cost value; therefore its selling value must shrink to the level of its reduced income. In this way a large share of the available assets of the small farmer, are swept away. The savings of years, are swallowed up and lost. Savings, that in the aggregate, amount to many millions of dollars. What has become of these values? They have been absorbed by the cities and the railroad monopolies, whose servants the cities are.

"Four decades of this process, has robbed the farm-center, as a unit of rural society, of its former wealth, independence and power. Rural society as a whole, is no stronger than its weakest unit. This is why agricultural districts are depopulated, while cities are over crowded. These results are the work of the competitive system, with its wasteful, wicked methods of distribution and exchange, which so widely separates the farm and the factory, the farmer and the artisan, the food and the consumer.

"From another point of view, we may discover that inventive genius, has added a long list of labor-saving machinery, to the equipment of the farm. Since wheat growing, has become the leading crop, this expensive machinery must be included in the outfit of every successful farm. The burden of this expense, has proved too great for the capacity of the small farm. It has encumbered thousands of them with an indebtedness so hopeless, that its annual interest swallows up the income of the farm. From these causes, a crisis in the affairs of agriculture has arisen, which has demanded larger farms, more capital, more brain force and more systematic, better organized, co-operative labor. Hence, the evolution of the bonanza farm ; with which the small farm can no longer compete. Notwithstanding its many wasteful methods, the bonanza farm has been a step in the right direction. It has taught our agricultural people a valuable lesson, as to what may be accomplished by the combined co-operation of brains, labor and capital. It has demonstrated the necessity for the evolution of the co-operative farm. It has prepared the way for it.

"With the advent of the co-operative farm, will come the beginning of a new agricultural era. The co-operative farm village, with its well organized, allied industries, will again unite agriculture with manufacture. The village will represent the new unit of rural society. This unit will be free, independent and self-sustaining The occupation of farming, will be lifted into a new realm. It will become the occupation of the noble, the cultured and the progressive. The people of these farm centers, will form the warp and woof of agricultural society, organized as a whole. The presence of organized society, largely adds to the value of all lands and to the value of agricultural and manufactured products.

"The brilliant author of 'Volney's Ruins,' well understood the force of this principle as applied to increasing agricultural wealth, and at the same time largely adding to the general prosperity of the State. In an essay published in 1790, Volney lays down the following principles: 'The force of a State is in proportion to its population; population is in proportion to plenty; plenty is in proportion to tillage; and tillage, to personal and immediate interest, that is to the spirit of property. Whence it follows, that the nearer the cultivator approaches the passive condition of a mercenary, the less industry and activity are to be expected from him; and, on the other hand, the nearer he is to the condition of a free and entire proprietor, the more extension he gives to his own forces, to the produce of his lands, and to the general prosperity of the State.'

"Each co-operative farm, will become a new center of permanent wealth; a new center of social progress; of organized labor; of distribution and exchange. These new centers, by again bringing together the food and the consumer, will save millions for themselves, which under the competitive system, were thrown away in freights and commissions. As these farm centers continue to increase, they may stretch away in one unbroken chain, perhaps five hundred miles in length. Each link in the chain, will be a five or ten-mile boulevard. Altogether, forming one continuous system of broad, free highways, the finest the world ever saw! Aided by trains of horseless carriages, there will be developed between the centers along this highway, a new system of transportation, distribution, commerce and exchange. With the establishment of each new system, the co-operative movement will gain an added impetus. The centers of exchange, distribution and commerce, located in great cities, will gradually lose their dominancy. The long lines of monopolized railroads, connecting these cities, will as surely lose a large proportion of their traffic. The magnetic wealth and bustle of the great city, will lose its attractive power. As a consequence, and by the action of a natural law, the tide of wealth and population, will flow back to the country; with its meadows and fields, its mountains and streams, its sunshine, blue skies, pure air and wholesome, enjoyable village life. Amid such surroundings, upright and just, fearless and free, the model citizen of a true republic, may find a natural home."

"Pardon me, Fillmore, for the interruption! I freely concede the desirability of the results, which you have so glowingly pictured. Nevertheless, I cannot quite agree with you, about the existence of a law, through which the tide of wealth and population will again flow towards the country. I am inclined to think, that facts and figures are against such a result. The statistics of the census of 1890, indicate that about one-third of the population, and over seventy-five per cent of the wealth of the nation, were then located in the cities. A little later, able thinkers and writers of the Josiah Strong type, proclaimed, that by the middle of the twentieth century, this would be a nation of cities, with less than ten per cent of its wealth and population remaining rural. As startling as these predictions are, I very much fear, that the logic of events favor their fulfillment!"

"If you will give me a little more time George, I think I shall be able to show you where these writers erred, in reasoning from wrong premises. They have judged the trend of events and the probable results that are to follow, from the standpoint of the competitive system. A system, which they have accepted without question as a permanent one, never to be replaced by another. This was the fatal error, which has robbed their conclusions of all value.

"In discussing the status of our great cities, these

writers all agree, that they are a constant menace to the nation; centers of political corruption, which are in every way antagonistic to the letter and spirit of a republican form of government; aggregations of the most dangerous elements of society, which are incapable of self-government. These admissions have a wonderful significance. Let us examine them.

"The question of society, becomes a potent factor in the solution of this problem. Society, like a great leviathan, covers the face of our country. Representing the aggregate of life, it affects all lives. As the social side of the body politic, it has the power to strangle or to nourish, every interest which is dear to those lives. Dominant society, is the support and inspiration of government. The excellence of any government, may be measured by the excellence of the society upon which that government is based. Under the standard of a republic, society may be divided into two classes; the true and the false. Reasoning from these premises, we may conclude, that in order to have a true republic, we must first evolve a true society.

"The society representing the competitive system, has its centers or units in our great cities. Its votaries, aro worshippers of wealth. They are importers of foreign fashions, and foreign ideas of government. They believe in caste. They detest equality. They have no love and very little respect for the equal rights guaranteed by the Constitution. They despise honest labor. They consider it menial, as a badge of servitude. They believe that wealth is a power which can raise the wealthy few to the dominancy of a privileged class. They believe that as members of this class, they can treat all other classes as servitors and dependents, who may be hired to do anything for money. They view with complacency, the crowded populations of our great cities. The greater and more dense the mass of people, the larger, more dependent and more obsequious the class of servitors. They are naturally, more or less in sympathy with monarchial and despotic institutions. They believe that the rulers, judges and lawmakers, should come from the ranks of the privileged class. They are out of harmony with the republic, because it is the true form of a co-operative government, Co-operation, they hate, it smacks of equality! They are devoted to the competitive system. They recognize its power to maintain a perpetual warfare among competitors, which shall forever keep the main host in such abject poverty, that they willingly become slaves to the wealthy. Having lost their independence, the votes of these competitors are at the command of their financial masters. Than this, nothing could be more harmful to the welfare of a true republic.

"This form of urban society, is the flower of the competitive system. The tendency of this society is to so engender selfishness, and to so destroy patriotism, that a multi-millionaire of the William Waldorf Astor type, deliberately achieves the acme of shame, by renouncing his allegiance to a country to which he owes everything. He expatriates himself, and flies to the refuge of a monarchy, to escape the honest burden of a just taxation. A taxation based on an assessment of less than one-third the rate, which is applied to the average farmer of the republic. One example of such ignominy, ought to teach every patriot, that the true republic must be built on the solid foundation of a society and industrial system, which represents justice and equality.

"Let us now question the co-operative movement, with the purpose of ascertaining its fitness to become the base of a new society, and also the proper foundation for a true republic. In a society growing out of the co-operative system, as our rural and agricultural societies may now do. We find the conditions are reversed. Labor, is the badge of respectability. It is the title to an honorable independence. In such a society, both men and women are free. All are co-operators, none are servitors. No beggars! No caste! The units of a co-operative society, are sound and healthy to the core. Co-operation, insures self-employment. Self-employment brings freedom, ambition, independence, self-respect, leisure and education; with all the comforts and refinements of life. With these insured, the co-operator cannot be bought or corrupted by wealth. Each co-operator becomes a citizen, who without fear and without restraint, may speak, write and vote, in accordance with the highest dictates of conscience. A healthful degree of honorable, self-sustaining labor for all, is the key-note of this social organization. Men and women are placed on the same plane of equality, financially, socially and industrially. For woman, this is a matter of the utmost importance.

"Productive co-operative labor, crowns woman with a self-supporting, self-respecting independence, which emphasizes her freedom from every form of bondage. In this, we have a perfect demonstration of the power of labor to bless humanity. Progressive life and invigorating labor, go hand in hand. One is the complement of the other. Labor as naturally promotes grace, strength, virtue and long life; as idleness breeds helplessness, vice, disease and extinction. Here we discover the wisdom, and the universal application of nature's law of labor. This law demands, that women who wish to become mothers of a dominant race, and who desire to secure perpetuity and progress for that race, must take an active part in some useful, productive labor. If we consider the significance of this demand, we shall perceive, that any form of social or industrial organization which denies this right to woman, or which takes from her the opportunity, the necessity, or the desire to labor, becomes her worst enemy, a foe to humanity, that is conspiring to reduce her to the degredation of a helpless dependent, a mere parasite. In her declaration, that 'The human female parasite, is the most deadly microbe which can make its appearance on the surface of any social organism;' Olive Schreiner has summed up in one sentence, the grave danger from this source which threatens the race.

"The combined and marvelous effects of the co-operative system and society on the woman question, rightfully places that industrial and social system far above all others, in the choice of a secure basis for the foundation of a true republic. In fact, George! After carefully considering the bearings of the questions involved, I feel sure that you will heartily agree with me in the assertion, that co-operative society, is the very embodiment of even handed justice, in which the rights of all are considered. Furthermore, you will be willing to admit, that it teaches the value of labor, and how to discover its uses and abuses. In eliminating its abuses, it will appear, that true progress, is to so improve and increase the ease and attractiveness of all kinds of labor, that they can no longer be classed as toil, or even disagreeable tasks. This then, is the legitimate field of inventive genius. Success in this field is assured, because it is in harmony with all laws of progress. Every hardship, every difficulty and every danger, which is eliminated from physical labor, increases in the same proportion, the opportunity and the demand for mental labor. This demonstrates the action of nature's law of compensation, which in elevating the character of labor, maintains its quantity."

"Yes Fillmore, I am convinced! I am willing to admit the truth of the assertions, which you have made concerning co-operative society, as the result of the co-operative movement. No doubt, they are destined in the near future to supersede the competitive system and the city society which grew out of it. As I view the situation now, that time cannot come too quickly! Yet, there is one point which still puzzles me. It is in connection with the rapid improvement of labor saving agricultural machinery, which, as Josiah Strong says, will soon enable a few farmers to do all the farm work, forcing all other agriculturalists to seek employment in manufacturing cities. How can you answer that argument, from the co-operative standpoint?"

"That is a pertinent question George, to which co-operation can furnish many conclusive answers. Let us consider the significance, and the conclusiveness, of some of the following:

"Under the co-operative system, every new labor-saving machine applied to agriculture, means just so much added wealth for the farm colony. It affords that much additional income, for active workers; so much more money to swell the annuity fund, for the retired members; so much more cash capital, for the sinking fund, with which to purchase, and to retain the permanent control, of an ever-increasing series of co-operative farms, for the lasting benefit of their people. With co-operative genius to invent, and an abundance of capital with which to buy, the advent of any conceivable quantity of improved machinery on the co-operative farm, would only serve to increase the wealth, leisure and independence of the co-operators.

"Such well-conditioned people could not, under any circumstances, be forced to leave homes of luxury and refinement in the country, to become the working slaves of a manufacturing syndicate in the city. Indeed! Why should they? Why should these co-operators, or any one with the opportunity to become such, go to the city to accept an insufficient and uncertain wage; to be compelled to pay five prices for food, when a better and more abundant supply, could be raised on lands of their own, with less than one-half the exertion? Having good homes of their own, why should these people pay exorbitant rents to owners of tenement houses, for the poor privilege of living in stuffy rooms, choked with smoke and filth, and surrounded by the clatter, the strife, the poverty and the soul-wearing competition of the great city.

"Why should they rob their children of health and happiness, by depriving them of a natural birthright, healthful exercise, free access to the pure air, the bright sunshine, the blue sky and the unnumbered charms of country life, with its fascination of ever changing landscape, a picturesque mingling of verdure clad hills, green meadows, shady forests, clear lakes and bold mountains? Why should these children be compelled to live a cramped, unnatural life, confined to the narrow streets, poisoned both mentally and physically, by the foul air, disease, corruption, crime and misery of the densely populated city? Why should agriculturists, who are independent co-operative owners of the soil, humiliate themselves by joining the vast army of struggling competitors, who throng the already overcrowded labor market in our great cities? Why should they be eager to become the financial and political slaves of the leaders of the competitive system; the social autocrats, who form the society of the 'Four Hundred?' "

"Can a Josiah Strong answer these questions? No! Why not? Because, in blindly reasoning and writing from the competitive standpoint, he has quite overlooked the fact that agriculture is the base of all wealth. He has forgotten, that as a class, agricultural people who own the farming lands of the country, hold the key to the situation. Made conscious of their strength by co-operation, they are the most independent people living. They are in a position to dictate terms to all other classes. They cannot be forced to do anything, which they do not wish to do. In arriving at his conclusions, it seems quite probable that Josiah Strong has made the serious mistake of accepting as true, a very prevalent idea, that in due course of business, (competitive business) all lands everywhere, would belong to the city capitalist; therefore, that all farmers would then be tenants at will, who could be turned off the land at the caprice of the owner. In this fatal mistake, we discover the error which has vitiated all premises from which he has been reasoning.

"Thanks to the forceful lessons, taught by Henry George, to which our agricultural people have given two decades of careful study. They have learned, that free access to land, is absolutely necessary to a natural enjoyment of life. They have learned, that for this reason, those who own land are masters of those who do not. With a sturdy independence which should characterize all citizens of a true republic, they have an intense antipathy towards all forms of slavery. Determined to remain free; they have redoubled their efforts to possess, and to retain permanent control of lands, sufficient for themselves and their children. In this work, they have discovered that co-operation leads to perfect success.

"In answering other arguments advanced to show why the city should dominate the country, and therefore absorb its population; the question of rent plays an important part. It should be studied carefully. The law of rent, is an enigma to the poorer classes, upon whose necks its yoke presses as a grievous burden. They sweat and groan under the burden, but can discover no way of escape. They must be educated. They must know the cause, before they can learn to avoid the effect.

"Rent, is a legal harness which enables the capitalist who owns houses and lands, to bind needy people to do his work. Through the exactions of rent, he can compel these people who can least afford to do it, to pay his taxes, his interest on capital invested, his living expenses, his traveling expenses, his insurance and such wide margins of profit, as necessity, opportunity and favorable location, may allow him to take. Rent values, like land values and market values, are exponents of social organization. Human lives, enter into the equation of these values. The absence of people diminishes these values, the presence of people increases them. For this reason, rents are highest in great cities, lowest in the sparsely settled country, touching zero on lands occupied by nomads. Land values, are affected in the same way. This will give us a clue, to the transitory character of wealth composed of values. It will give us another reason, for the shrinkage in value of farm lands, and the increased wealth of cities; which follows the migration of people from country to city.

"We may now consider another important factor, which affects rent values in great cities. It is the spur of a sharp want, of the urgent necessity of helplessness, which must drive and control the actions of a large majority of the inhabitants. The presence of these elements is necessary, in order to create the highest markets for rents. The larger the throng and the keener the necessities of the crowd of bidders competing, the higher the prices they will pay for rent. Under the reign of the competitive system, this is a conclusive demonstration of the truth of the saying, 'That the necessities of the poor, are the opportunities of the rich.' Is anything further needed, to prove that the competitive system is the essence of a cruel barbarism, which blots the civilization and shames the humanity of the republic? Why not change it for the co-operative system?

"Under the progressive and beneficent reign of co-operation, there would be homes for the homeless, land for the landless, work for the unemployed and independence for all. This would mean, a total absence of want; that imperative spur, which is so necessary to the life of competition.

"Transportation and taxation, are two factors yet unnoticed, which materially affect rent values in great cities.

"Taking up the question of transportation; we

soon discover its importance. The great manufacturing city, is the center of a complete network of railroads. The inhabitants of the city, are at the mercy of these railroads. Nominally, they are supposed to be competing lines. As a matter of fact, by means of traffic association, they become one huge, consolidated monopoly. A monopoly so dangerous, so powerful, so unscrupulous, and so voracious, that it does not hesitate in fixing and maintaining rates so exorbitant, as to be actually prohibitory, at least so far as two-thirds of the city dwellers are concerned. Meanwhile the monopoly arbitrarily depresses rents and land values in the country, while it increases them in the city.

"Let me give you an illustration of the methods, by which these results are accomplished. Take if you please, the case of an average city, factory-worker; receiving an average wage of one dollar and fifty cents per day. On this wage, he has a family to support. In the country, thirty miles away, he can have a comfortable house, with a nice large garden, for the moderate rent of five dollars per month. A most desirable home! But, here comes the opportunity for the railroad! A ten cent fare each way, six days in the week, would pay the railroad a handsome profit. But, a handsome profit does not satisfy a monopoly! The handsome profit must be doubled six times, before it will consent to serve the public! As a result, this workman, not having the ready cash with which to purchase a monthly commutation ticket, must pay to the monopoly, at its lowest rate (two cents per mile) the gross amount of one dollar and twenty cents per day for transportation. Subtract this sum from the workman's daily wage; there will remain the scant trifle of thirty cents, with which to pay bills for food, fuel, clothing, medicine and other family expenses. Utterly impossible! Even if the owner of the country house and lot, should consent to reduce its price and its rent one-half, the workman would still be prohibited by the railroad, from taking advantage of the reduction. He would gladly pay the ten cent fare, for then he would be able to pay ten dollars per month rent, for the luxury of occupying such a desirable country home. This would be a blessing to all interested parties; still, it cannot be, because the monopoly says no! Being a monopoly under the protection of the competitive system, its dictates may not be questioned.

"Although, the case cited, may be duplicated a thousand times, every day in the week, in every large city of the republic; yet, everywhere, on all possible occasions, the common sense of the people is outraged, and their ears offended, by the loud shouts of the competitive leaders, who praise without stint the great usefulness of the monopolistic trust. Solemn as owls, with an air of great learning, they assure the people that these beneficent trusts, are the natural outgrowth of high-grade business methods, which must be let alone. Do the poor people, the farmers, the country land owners, and the working men, join in these shoutings? Obviously and most assuredly, they do not!

"Let us now follow our factory workman back to the city, for the purpose of noting the effect of this monopolized transportation, on city rents. Baffled in his desire to live in the country, he seeks to make the best of a bad situation. As a consequence, he is obliged to pay to the owner of some tenement house, a rental of fifteen dollars per month for three small rooms; poorly ventilated, unfurnished and unheated. These rooms are so undesirable on account of difficult access, bad location, unsavory smells, and the immediate presence of other tenants in the house, who are quarrelsome, drunken, filthy and generally disreputable; that but for the prohibitory tariff maintained by the railroads they would remain unoccupied, even if the rent should be reduced to seven dollars and fifty cents per month. However, poor workmen receiving scant wages, may not expect to be choosers. They with their wives and children, must ever bravely strive to adjust themselves to their environments, which more often than otherwise, prove cruelly bitter and oppressive.

"In the case of our artisan, who is a brave, industrious, hopeful fellow; after paying his rent, he will have left from his monthly wages, the small sum of twenty-one dollars. Providing of course, that throughout the month, he has been so fortunate as to remain well and to lose no time. With this amount, (seventy cents per day) he must manage as best he can, under such adverse circumstances, to feed, warm, clothe, shoe, and protect his family. With such a meagre sum to supply so many wants, it is impossible for him, even under the most favorable circumstances, to make petty savings with which to meet emergencies. When the misfortune of sickness overtakes him, the situation becomes appalling!

"From this illustration, we may judge how much the city is indebted to the railroad monopoly for its high rents. To great cities, high rent is a matter of the utmost importance. Take all rent advantages from them, and the entire list of their manufacturing industries, could be carried on in country villages with equal profit. It is quite evident then, that these cities are alive to the fact that rent is a measure of the value of locations."

"Before going farther, Fillmore, allow me to inquire! Why could not these working men and their families, who are confined to the city by the high rates of the railroad monopoly, find cheap country homes near the city; say within a radius of from five to ten miles?"

"Thank you George, for such an opportune question! Its answer leads directly to a discussion of the question of taxation.

"A land monopoly, is more to be feared, more harmful to the poor and more disastrous to the interests of the general public, than any other kind. The worst form of land monopoly, may be found in full force, along the outskirts of large cities. These monopolies are made possible, by the unjust application of a faulty system of taxation.

"As a preliminary step, a hungry host of individual capitalists and land syndicates, proceed to purchase large tracts of adjacent lands at farm prices. These lands are then sub-divided into villa sites, and into a variety of sizes of town lots. Prices are placed on these lots, which would about equal the value of the ground, when in course of time, at the edge of the city, they should be covered by dwellings or business houses. This accomplished, the holders like cormorants, sit and wait for the growth of the city and the efforts and capital of other people, to so increase the value of their holdings, that they can realize their prices and take their profits. These periods of waiting, may cover a long time, often, from one to twenty years. Meanwhile, these monopolized lands are kept out of use, because on account of high price, they cannot be used for agricultural purposes.

"Why can these land monopolists afford to wait so long? Because an inequitable system of taxation, discriminates in their favor; offering aid and encouragement for them to do so. Without this aid, it would be impossible to keep these lands out of use.

"How can this happen? In the first place, these sub-divided lands, as a whole in large tracts, are assessed at the rural rates applied to unused and unoccupied lands. These assessed values, may be so low, as to be less than one per cent of the asking price of the lots. As time passes, they are liable to be slowly increased. Under such a discriminating system of assessment, the taxes that may be collected, are merely nominal. This unequal system of taxation, is applied, in a proportionate degree, to all unoccupied lands inside the city limits, which are held out of use by the land speculators.

"How does this state of affairs affect city rents, and at the same time, assist in preventing the poorer classes from enjoying the advantage of country homes? First, it establishes a broad zone of monopolized land around the city. This zone continues to increase in width with the growth of the city. Scattered through this zone, are many tracts of farming lands in active use. For this reason, they have to bear an extra burden of taxes, in order to equalize the low rates on such large tracts of idle land. These heavy taxes are patiently borne by the resident farmers, with the hope of reimbursement in the near future, by being able to sell their farms for extraordinary prices. In this way, abnormal prices become firmly established throughout the zone; which like some great barrier most effectively confines the working man and his family, to the narrow limits of a city tenement, with its high rents.

"If a builder with some idle capital, should wish to erect a considerable number of modest cottages, within the limits of this monopolized zone; with the purpose of renting them to working men; he would find it impossible, or at least impractible to do so. Why? Because he would have to pay almost city prices for the ground; then, having covered the lots with houses, he would be obliged to pay a heavy penalty for this outlay of capital, by the grievous burden of taxation, which would fall upon him. Houses built under these circumstances, could not be let at a rent low enough to be within the means of the working man.

"The number of people who are confined to city life by the causes named, is very large Just how large, I have no means of ascertaining. Families, who are subsisting on incomes of ten dollars per week and less, furnish a large proportion of this number.

"We have seen that the disastrous crowding, the alarming density of our large city populations, is mainly due to two causes. High transportation, caused by the railroad combine; and an outrageous land monopoly, made possible by a bad system of taxation. We have seen, that this dense mass of needy humanity, constantly creates such a fierce competition, that rents must grow higher and wages must grow lower. We have seen, that the causes named, are steadily diminishing the wealth of rural sections, by transferring it to the great city. We have seen that this whole movement, which tends to transform the great majority of the independent citizens of a republic, into the financial slaves of an oligarchy, is the natural outgrowth of the competitive system. Taught by history, we know, that as the oligarchy rises and reigns, the republic dies.

"Knowing the causes which have produced these conditions, we are prepared to discover, and to apply the most efficient remedies. It is only by associated effort, that rural populations can successfully oppose the concentration of wealth in cities. The well organized mass, becomes a great power. The new century demands a new industrial organization. The co-operative system, answers the demand. It is in harmony with the idea, that life is the most precious of all things. Therefore, it recognizes that opportunity to labor, and to enjoy the fruits of that labor, is the highest privilege of life. Under the reign of co-operation, this is insured. United in congenial co-operative associations, farming and working people in the country, reinforced by large numbers of recruits from cities, may build up for themselves, new centers of combined industries, society, wealth, distribution, exchange, education, amusement and insurance; which will place them in the ranks of the self-employed, who are financially and politically free. By growth and expansion, these centers will become the units of a vast co-operative system, which must soon wholly displace the competitive.

"The inspiring motive of this co-operative system, will be the elevation and perfection of human lives. To this end will tend the invention of every labor-saving machine; increasing the product and shortening the hours of labor. With the physical man thus properly nourished and developed; the intellectual and spiritual man, will for the first time in history, have the necessary conditions in which to expand, blossom and bear fruit. Under such circumstances, life in the country will be both altruistic and idealistic. By comparison, life in cities will become a hardship which few will care to choose. The few, it may be taken for granted, will be so bound to the wheels of Mammon that they cannot get away.

"The larger independence and better education of the co-operative majority of voters, will soon enable them to find a relief for the imprisoned populations of cities, which are now confined by the pressure of land monopolies and railroad combines. They will see to it, that these railroads become the property of the government; well knowing that they can never be made to serve the public honestly, until the public owns them. As for the land monopolists, they will find their holdings so burdened with taxes, that they can no longer keep them out of use. The erection of fine buildings will be encouraged. Costly mansions, dwellings, or factories, will not increase the tax. With these barriers removed, the densely packed populations will quickly expand. They will fly from center to circumference of the city. Later, they will be attracted to the country village, where more congenial homes and employments await them. Then educated and emancipated, they will no longer pay rent.

"We have seen that the economics of society vitally affect the status of human lives; physically, morally and spiritually; industrially, financially and politically.

"We have seen, that rural society, based on the co-operative farm colony as a unit; answers every demand for the protection and development of human life. We have seen that the inspiration of this society, is to secure for all, a lasting reign of peace, plenty, harmony and progress; a most convincing proof, that it is the ideal society on which to build a true republic, that shall be self-sustaining.

"We have seen that the perfect emancipation of woman, and the exalted motherhood, which is made possible by the advantages of the co-operative system, insures the permanency and the dominancy of a republic so supported.

"In analyzing the workings of the competitive system, we have seen that its methods are those of war. In the never-ending struggle of competing strife, opposing armies of human beings slowly grind each other to death; leaving unaccomplished the real object and purpose of life. This enormous waste of life, violates every principle of a republican form of government. It aborts even the efforts of planetary evolution.

"We have seen that the competitive system produces monopolies and trusts, with a constantly increasing tendency to concentrate wealth in cities; placing it in the hands of the few, who are the financial masters of the many.

"We have seen that from the ranks of the wealthy few, come the leaders of competitive society, who make their strong holds in the great city. They are the shining lights of the competitive system. They believe in a constant warfare of competition, which brings suffering to the many and success to the few. We have seen that a surfeit of wealth and power, has made these leaders so despicably selfish and unpatriotic, that they are unwilling to pay a just proportion of tax for support of the government.

"We have seen that the monopolist, encouraged by the sympathy of competitive society, endeavors to monopolize administrative and executive functions. By means of unequal rates of taxation, and more especially of unjust assessments, he is able to shift most of his taxes to the shoulders of farmers and small property holders in state, county and town. This outrageous evasion by the rich, of their just share of the burdens of government, is shameful to the last degree! It robs the poor of all protection, that governments are bound to offer! It is a crime against humanity! It is a sin against the perpetuity of the republic! It is anarchy! If a government is no longer able to protect its poor; then, such a government has forfeited all right to exist!

"We have seen that a true government, republican in form, is a co-operative institution, which must be based on justice, and equal rights, for all; thus recognizing the common brotherhood of humanity. Organized and maintained for the purpose of conserving, developing and protecting life; such a government, would at all times be guided by the beacon light of the axiom, 'That the injury of one is the concern of all.' It would wisely measure its strength and perfection as a government, by the strength and perfection of its weakest unit.

"We have seen that with members of competitive society, the accumulation of wealth, becomes the sole ambition of life; that they may enjoy the ease, luxury and social power which follows. We have seen that wealth develops selfishness and idleness. Idleness breeds helplessness, vice, disease, and extinction. The predominance of such a society, would mean the death of the republic.

"Having compared the merits and demerits of the two industrial systems, and of their closely related societies; taking it for granted, that as the highest expression of social evolution, the republic must endure; which, George, do you think will prove the true system, the true society, that must predominate;

that must naturally develop most social and political power; most perfect conditions of life; most happiness?"

"There can be but one answer, Fillmore! The co-operative is the true system, and the true society! You have made it very plain that the republic cannot endure without them. It is equally evident, that with restraining influences removed, city populations in a large measure, will again return to the country for homes; attracted thither by the many advantages offered by co-operative village life."

"Speaking of homes, George, reminds me that I must now confer with you in regard to a personal matter, which may affect your work and your welfare for many years. This is the fifteenth of September. You have now been in Solaris, a little over one month, with an opportunity to study the co-operative movement quite extensively. I believe you are in harmony with it; and can do a good work for it.

"This office, as you know, is the present headquarters of the general movement. Tomorrow I am going East, to be absent at least one month, perhaps three. I wish you, as my private secretary, to at once take charge of the office. I can offer you a salary of $1,500 for the first year. The office staff is a capable one, which will make your work quite light. I have made arrangements with Mr. snd Mrs. Gerrish and with Miss Houghton, to co-operate with you as advisers. Since the first establishment of the office, Miss Houghton has so often volunteered to assist me, that she is now familiar with the routine work. Finally, I shall at all times while away, be within reach by phone or wire; by which I wish you to consult me whenever occasion may demand. What say you, George! Can you accept my proposal?"

"Yes, Fillmore, I accept without one moment's hesitation! I shall be delighted with the opportunity to work for the interests of co-operation. You may trust me to do my best!

"By the way, Fillmore! I take it for granted, that before you return you will meet Miss Fenwick, and her friend Mrs. Bainbridge, if so, please present my regards."

"I shall not forget your message, friend Gaylord! Miss Fenwick is now at Fairy-Fern-Cottage, on the Hudson. She will meet me at Fenwick Hall, in Washington, where we are to be married on the twentieth day of this month.

"The wedding is to be strictly private and informal, only Miss Fenwick's attorneys are to be present as the necessary witnesses. After the wedding, the customary tour will be omitted; leaving us free to remain at Fenwick Hall, until the inspiration of the moment brings the choice of some mountain or sea-side resort.

"I shall expect you, George, to mail weekly reports from the office, to Fenwick Hall. Wire me for instructions, whenever you are in doubt."

"I shall obey your wishes to the letter, Fillmore! What you tell me of the coming wedding, is glorious news! I congratulate you with all my heart, on your great good fortune! You deserve it; you have well earned it!"

CHAPTER XLIII.

A TWENTIETH CENTURY HONEYMOON.

AT FENWICK HALL, in the early twilight of their wedding day, we find our hero and heroine, the bride and groom, now husband and wife. They are sitting side by side, hand in hand, looking forth from the large southern window of that magnificent tower room, hitherto known as the private retreat of Fern Fenwick. The outlook from that window was a revelation of beauty, as perfect as a dream of fairy land.

As the twilight deepened, high in the southern sky, the full-orbed splendor of a September moon, glorified with its soft radiance, the marked beauty of the Capital City—the Pearl City of the republic. From the mysterious depths of stilly night, intensifying the soothing charm of moonlight; there came softly stealing through the open window, the balmy airs of evening, laden with the fragrant breath of a thousand flowers. From the Aqueduct Bridge to Fort Foote, a long line of brilliant light, with many a graceful curve, marked the pathway of the broad Potomac, whose unruffled bosom shone like a mirror of burnished silver. Stretching across the valley from distant heights, a fleecy veil of enchantment woven in the loom of mist, etherealized city and river, dome and monument, tower and steeple, cottage and castle; adding a weird beauty to the magnificent array of public buildings, which owned the Capitol and the Library as chief. Above and beyond all else in its unapproachable glory, the Dome of the Capitol in the mellow, hazy moonlight, shone resplendent as a matchless crown to the architecture of the Occident!

Responsive to the spell woven by the fairy fingers of moonlight, in which soul and sense sink to the spiritual repose of that serene calm, where in silence, happiness of the purest type best expresses itself; these newly wedded lovers, living in the inner world, lost to the outer, remained motionless and absorbed in the ecstasy of contemplation.

Fern was the first to break the silence. She said: "My dear Fillmore! Tell me, is this the beginning of some reign of enchantment? The culmination of love's dream? Are we waking or dreaming? Can it be possible, that this glorious moonlight, so auspiciously ushering in our honeymoon, is typical and indicative of its endurance, of its unalloyed brightness?"

"My wife! Chosen one of all women! Your devoted lover for six years; having passed the stage of love at first sight, hopeless love, worshiping love from afar, patient love, love requited and love rewarded; I am now so happy, so unspeakably optimistic, that I accept without question the happy augury of enchanted moonlight, as being truly prophetic. Besides, having a wife so noble, so good and so wise, to make it possible; how could our honeymoon be other than the most delightful ever known to the history of love? You may trust me, dear heart, to do my best towards making that prophecy come true!"

"In discussing honeymoons, even my own; I may not be permitted to trust, in what is given to me to know. As a maiden of twenty-six summers, now your wife; I know very well that a husband who is just, loving, noble and true, is the most important of all factors, in securing the perfection of the ideal honeymoon. That six-year ordeal of loyal, patient love, which you have so thoughtfully analyzed and classified, has made you very dear to me! In overcoming this ordeal so victoriously, you have displayed a strength of character which has commanded my admiration. You have been unselfish, courageous, persistent of purpose, trustful, thoughtfully sagacious, perfectly trustworthy, and strictly honorable. For these characteristics, so like those possessed by my father; I love you more than for all else. Since crowned with conscious life, my father has been to me, the standard of an ideal man! If ever a daughter worshipped a father; I was that daughter. In character, you, of all the men I have met, are the nearest like him. Stronger words of praise than these, the lips of a proud, loving wife, could not utter! Now Fillmore! My dear husband! I am going to kiss you, as an antidote; lest the fervor of my speech, should make you vain, just a little!"

"The antidote seems to work like a charm! Yet, a speech so full of such crushing praise, coming from the lips of the loveliest and most thoughtful of wives, is very provocative to vanity. It makes my case so desperate, that it really requires heroic treatment. To make the antidote effective, I should say, increase the quantity of the dose administer very frequently!

"But seriously, my dear wife! I am overwhelmed by the tribute of praise, which you have paid to my character! To me, the character of Fennimore Fenwick, is nobleness personified! To have my own continually compared with one so exalted, is a very trying ordeal. I tremble for the consequences! I am now so happy, that in the very selfishness of my love for you, I may shatter your ideal. To disappoint you; would be to forfeit my paradise! In times of trial, I shall appeal to you as the noblest and best of wives, to use your highest gifts of occult power to assist me in retaining your respect, admiration and love. Meanwhile, my dear wife! I shall cherish in my heart, the memory of your tribute, as a talisman, as a perpetual inspiration to live up to my highest ideal! Whatever happens, I shall be myself."

"That, Fillmore, has the true ring of your natural nobility! Be yourself, and we shall be lovers forever! With that question settled; under the inspiration of this lovely moon, let us commence the construction of our castles in the air. In marrying a woman with a great fortune, you have pledged yourself to share equally with her, the pleasures, cares and responsibilities of her riches. Remembering, that henceforth, we are joint trustees, under my father's direction, for the wise use and distribution of this wealth. It becomes our duty to make competent and well-considered plans for the work. What say you, my dear husband! Shall we not do well, if we devote a generous share of our honeymoon to the making, development and perfection of these plans?"

"What you propose, my dear Fern, will make me very happy! I shall be delighted with the opportunity to relieve you of a portion of the burden of your responsibilities, by sharing them. How, and when shall we commence the plan making?"

"Before undertaking the plans, it will be necessary for us to ascertain just how much we are worth, financially speaking. For this purpose, we must make a complete and carefully classified inventory of our properties, both real and personal. This important task, we will take up tomorrow, working deliberately until it is finished. It is quite likely to prove a long one, bristling with interesting data, suggestive and educative, as to the extent of your newly assumed responsibilities.

"After the inventory is complete, we will each in favor of the other, make and execute a will, conveying the property described by the inventory. Then, we shall be prepared for the accidents, emergencies and unexpected changes of a mortal existence.

"Having disposed of the wills, we will return to the inventory. Going over it without haste, item by item. While considering each one, I will give its history; then, we will make a short note, embodying our individual ideas as to the best present or future disposition of that particular piece of property. These notes to be attached to the inventory. By the time we have finished this work, you will have acquired such a firm mental grasp of our financial situation, that you can advise me wisely, or act alone, as the occasion may demand."

"Pardon me, sweetheart! What of our coming conference with your father, Fennimore Fenwick? Is that to be postponed until we have finished the preliminary work, which you have outlined?"

"Yes, my lover! I would not have you take part in the consultation, without first being equipped with this important knowledge. Besides, it was so understood, by father and myself, when we arranged to have the conference take place on the afternoon of the fifth day after the wedding. There will be plenty of time. You are perfectly satisfied with the arrangement, are you not?"

"More than satisfied, my good angel! I can

hardly realize my good fortune! I am eager to begin the work. What a delightful time we shall have! To have you introduce me to our wealth, by the way of this unique, honeymoon program; is something very like a fairy story! I could not devise or imagine anything more delightful!

"Six years ago, at the time of our meeting, I was hopeful and ambitious. My heart was filled with an earnest longing for the fulfillment of my one great purpose in life. But, how to accomplish that purpose, was hidden from me by the veil of the future. Then, I never dreamed that waiting behind the veil, love was the goddess of good fortune, who was to guide me to success! It is the unexpected which always happens! Thinking not of self; destiny smiled on my unselfishness, and kindly led me to my fate! Having met you, I dared to love! Discovering that you cherished a purpose in life like my own, I dared to hope! Trusting to love, as the messenger of destiny; in the unalloyed happiness of this glorious honeymoon, I have reached the goal of all my ambitious hopes! When I reflect on the magical change of my environments, and the new career in life which has opened for me; I can appreciate the full significance of the miracle which love has wrought!

"Knowing the importance of unselfishness on the part of the individual, as a necessary factor in the successful co-operation of the multitude; I perceive that selfishness must be overcome by a comprehensive system of education, organized for that particular purpose. The organization of such a system must be accomplished by a small number of enthusiasts, who are willing to devote their lives to it. This means, that they must be people of wealth and leisure.

"As an evidence of appreciation of responsibility, for my stewardship of the wealth which you have bestowed upon me; I wish now to declare my purpose. It is, to devote the remainder of my life to this educational work. It now comes to me, that this is the work described for us, in your letter, written to me over thirty months ago; where, in a vision of the future, you saw us united, side by side, hand in hand, fighting successfully against the poverty breeding hosts of selfishness. From the innermost depths of my being, I rejoice over this most fortunate opportunity, which permits me to take an active part in such an important work! My heart swells with pride and happiness, when I feel and know that I am to have the honor of standing by your side, in the fore-front of the fight!

"I can now appreciate the utility of my long apprenticeship on the co-operative farm. In no other way, could I have been so well prepared for leadership in the educational movement. I have learned just what agricultural people need to make them perfect citizens of a perfected republic A republic of peace, without a police; without the burden of a standing army, to menace and oppress its citizens, because they are already a law unto themselves, at peace with all the world. When I analyze the influences which have inspired and led me, throughout this extraordinary course of training; I recognize the action of a dominant, guiding mind; the far-seeing wisdom of my noble friend and benefactor, Fennimore Fenwick. To him, and to the spirit world, I shall ever be profoundly grateful! Is it not a most beautiful illustration, of the power of spirits to co-operate with mortals?"

"Very true and rightly spoken, my prince of

husbands! I too, am glad, that during the six years of your preparatory training, destiny's messenger —love—has guided you so wisely. With your intuitive nature, I am not surprised that you have divined so clearly, the general scope of the life work, which my father has planned for us. At the coming conference, he is to unfold the details of the work. Let us well employ the intervening time, in doing the preliminary work; which, as you have so well said, will give us an added relish for the enjoyment of our delightful honeymoon."

CHAPTER XLIV.

THE NEW CRUSADE.

THE BEAUTIFUL seance room at Fenwick Hall, was known to the chosen few, as the "Tower of the Psychics." In fittings, furniture, and equipment, it was much the same as the square room in the central tower at Fairy Fern Cottage. From the beginning, this room had been devoted to but one purpose; that of an audience chamber for the intercommunion of the Two Worlds, the spirit and the mortal. Every visiting mortal felt the presence of a refined spiritual atmosphere, a highly charged, electrostatic potential, which made possible superior spiritual conditions. In this room, Fennimore Fenwick was at home, to the chosen few of his friends on the mortal plane of existence. On the afternoon of the conference, we find our hero and heroine in this room, awaiting the coming of Fennimore Fenwick.

While Fillmore was admiring the full length, life

size painting of his spiritual friend and benefactor, which hung on the wall opposite the entrance to the room; the familiar voice of the original, through the trumpet very near, gave him a cordial greeting.

"Bless you, my son! How glad I am, to welcome you to Fenwick Hall, as its new master! May your reign here as such, prove long and prosperous! In the enthusiasm of my fatherly pride, allow me to congratulate you on your rare good fortune, in winning the hand and heart of my daughter, Fern. She is a pearl above price! Ever love her devotedly, my boy! Cherish her tenderly, as the brightest jewel in your crown of life!"

"Thank you, Mr. Fenwick! For your affectionate and kindly words of welcome! To me, they are more gracious, more inspiring and more delightful, than words can express! They have so taken me by surprise, that I am overwhelmed by the strong tide of emotions welling up from my grateful heart! As to your commands in relation to my precious wife; you may trust me! Waking or sleeping, I shall never forget them! They are burned into my heart, by the intensity of my love for her, by the force of my lasting esteem and admiration for you! How can I ever properly thank you, my noble benefactor, for your great goodness to me; for your supreme confidence in my integrity? In return, I can only ask you to accept my pledge, to ever strive to merit that confidence!"

"Do not thank me, my son! Thank Love! Destiny's messenger; who as a reward for your unselfishness, has kindly led you to the goal of your present happiness!"

"And you, my beloved daughter! Are you quite happy! May I also congratulate you, on having so wisely chosen a husband, who is in every way worthy? Do you remember the promise I made to you, on the night of my transition? A promise to bring to your side, a friend, a counselor, a protector, whose wisdom and integrity, should at all times, prove sufficient for the needs of the hour. Are you satisfied, my dear girl? Have I faithfully kept my promise?"

"Yes, father! I am more than satisfied! I am a contented woman, I am very happy! The quiet delicious calm of my happiness, is a new experience for me. Heretofore, I had supposed that happy women must be vivacious and voluble, from the very effervescence of their happiness. Now I know that it is not so. Your characteristic words of praise, for the one I have chosen as a husband, have made me very proud of him and deeply grateful to you! In him, I have found the promised friend, counselor and protector; also, an ideal lover. But, my dearest, kindest, best of fathers; you know very well, that to trust you implicitly, is a law of my life! I have always trusted you! Therefore, I am not disappointed; neither am I very much surprised. I am just perfectly happy. That is the whole story in a nutshell!"

"This is as it should be, my children! When I first saw you, Fillmore, I felt intuitively, that you and Fern were made for each other. I knew I could trust you together, to finish my work. Now, I rejoice, that my intuitions were so prophetic!

"In your work at Solaris Farm, Fillmore, you have succeeded beyond my most sanguine hopes. I congratulate you heartily, my son, on this initial success for the co-operative movement! This is but the beginning of the work. As we go farther, wider fields are opened for more extended efforts.

You have already correctly surmised, that selfishness in humanity has become so dominant, so crystallized, from long centuries under the heartless reign of competition, that only a far-reaching, well organized, especially designed scheme of education, can conquer the evil. By means of this educational program, we shall be able to open the eyes of both poor and rich, to the benefits of co-operation.

"It has been wisely and truthfully said, that: 'The destruction of the poor, is their poverty. That conversely, the poverty of the poor, is the real power of the rich.' In these two short sentences, we have the most scathing indictment against present social and industrial conditions, that could be made! These conditions are wickedly abnormal! They are entirely out of harmony with the law of progress, and of planetary evolution! To change them for something better, is the crying need of the hour!

"It were a mercy to both rich and poor alike, to make them financially independent of each other! Then, freed from the thraldom of selfishness, they could discover and appreciate, each for themselves, the true object and purpose of human life. For this reason, our new educational movement, must be so arranged, that it may successfully appeal to all classes.

"For the industrial classes, the agriculturalists and the artisans, we can use the co-operative farm movement as a basis of education. As for the wealthy remainder, they must first be taught to respect the sacredness and the true purpose of human life, before they can contemplate any form of social or co-operative progress, with feelings other than contempt, or at least angry opposition. This is to be expected. It is the natural outgrowth of the teachings of a society, which is controlled by the hierarchy of competition. Both the co-operative farm and the broader educational movement, are to be embraced by the work of the New Crusade.

"The New Crusade, is to be organized, promoted and maintained, for the peaceful conquest of poverty; and the consequent banishment of ignorance and crime. These grand purposes, shall be emblazoned on its banners, appealing to the chivalry and knighthood of the republic for support. Never before has the bugle of the crusader, blown the assembly call for so noble a cause! Victory for this glorious cause, means a recognition of the true nobility of labor: The establishment of peace on earth, and happiness for all: An abundant harvest, for all productive toil: The sacredness and divine significance of life: The brotherhood of humanity: And the solidarity of all social interests. To the victors, shall come the well earned plaudits of a thousand future generations; whose sons and daughters shall chant the story of the unparalleled chivalry of such noble, unselfish deeds!

"To you, my children, is assigned the task and the honor of inaugurating this peaceful campaign. From you, it will demand extraordinary activity, courage and administrative ability; reinforced by large sums of money. Fortunately, the Fenwick fortune is ample. Use it without stint. Fenwick Hall, is roomy and well fitted for the headquarters of the New Crusade; and for the housing of its organizing staff; which, from the magnitude of the work, will be a large one. A bureau of literature must be formed. A newspaper and a magazine, devoted to the cause of the Crusade, must be published. They must be the best of their kind. The editorial talent must be of the highest order, the ablest in the land. Every State in the Republic, must be made a department of the Crusade. A select army corps of teachers, organizers and leaders, must be assembled, trained and thoroughly prepared, to take charge of these departments. They will be the executive and recruiting officers of the Crusade; rendering weekly reports to the headquarters in Washington. Every co-operative farm, will become an outpost and a recruiting station; every State, a grand encampment.

"In recruiting crusaders from the ranks of the wealthy, a special effort should be made, to have them take up the cause as a fashionable fad. They can be diplomatically led, where they cannot be coaxed or driven. In the face of any opposition they may display, it must ever be borne in mind, that the hearts of nine-tenths of the wealthy, are good and true. Their natural promptings are to do right; to use their riches for the advancement of science, and for the cause of humanity. They would do better, if they only knew how. They must be educated. The competitive system, under which they were born, trained and made rich, is at fault. By it, they have been taught, that poverty is a necessary and permanent state; to which, a large majority of the people of the earth, are assigned by the action of a divine law. Therefore, any attempt to banish poverty would be not only useless, but actually sinful. Nevertheless, prompted by a higher law, many of them annually dispense large sums in charity. Under the competitive system, charity only aggravates the malady. It is money thrown away! As the recipients are thus enabled to work for less wages; increasing the gains of competitive masters; and finally, swelling the ranks of the helpless poor. After a few trials, even the most persistent alms-giver soon discovers, that as an antidote to poverty, charity is a wretched failure. Taking it for granted, that the competitive system is a permanent one which is to endure forever, he gives up the problem as hopeless.

"It is to be the business of the New Crusade, to show why the co-operative should be substituted for the competitive system. It must teach the wealthy classes, the vast importance of the great lesson taught at Solaris. Namely, that by organized, unselfish co-operation; independent self-employment, producing an abundance for all, may be speedily and practicably substituted for every form of poverty. The Crusade must demonstrate, that ignorance, poverty and crime, are handmaidens, which cannot exist apart. That if one-half the money expended for charity during the past fifty years, had been used to promote co-operative self-employment, poverty, tramps and ignorance, would now be things of the past.

"To the people of the republic at large, must be taught the significance of the contrast between the war-like competitive system, and the peaceful methods of a co-operative association. Co-operation, makes combined individual effort, equal to the wealth of independence. The co-operator, being self-employed, no longer strives to displace a fellow workman by offering service at a lower price.

"Competition, emphasizes the poverty and helplessness of the individual, because it sets every man against his neighbor, against the whole world. The competitor deliberately shuts himself away from all gain that might come to him from the force and effectiveness of associated effort. He loses all faith in mankind; in honesty and justice. He views the good fortune of a fellow toiler, as a personal injury, which he ought to resent. In fact, he becomes too selfish to even be patriotic!

"The quickest way to convince the people of the barbarism, the cruelty, and the wickedness of such a system, is to establish a co-operative farm in every available township throughout the land. The free, healthy, trained, and well-educated social communities, growing up on these farms, will become the units of a true society; the underlying foundation, on which to build the true republic.

"Society dominates the political expression of nations. It molds and controls public opinion, business methods and commercial usage. Under the reign of competitive business and society, the market is largely composed of small wage earners, whose necessities are so great, whose tenure of employment is so uncertain, and whose wages are so scanty; that they are forced to buy the cheapest of everything. On the part of tradespeople, the fierce competition to control this cheap market, encourages the use of an outrageous system of food adulteration, and with it, every possible degree of lying, cheating, fraud and deception; until the moral tone of both business and society, has become blunted; yes, well nigh destroyed. As a result of this shameful state of commercial affairs, the successful man in any line of business, can no longer afford to be honest. He knows very well, that in competitive business, he can utterly ignore honor, conscience, and self-respect, without losing the approval of competitive society. Can such a rotten society ever become a safe foundation for the government of a true republic?

"It is to be the mission of the New Crusade to teach and to demonstrate, that under the reign of a co-operative system, and society, these conditions would be reversed. All incentives to cheapen goods, or to adulterate food products, would vanish. The co-operators would then form the bulk of the market. Buying at wholesale collectively, to sell to themselves individually; they would be in a financial condition to pay remunerative prices, for whatever was genuine, pure, wholesome, good, reliable and lasting. Inferior articles, they would not purchase at any price. The demand for cheap stuff would cease. The dominant motive of the commercial world, would be revolutionized. Among manufacturers and producers, the cry would be, not how cheap, but how excellent, can we make our goods! The long-practiced, skillful chicanery of competitive methods, would be at a discount; they would be worse than useless! Honest men could then engage in business, without violating either honor, or conscience! Cheating and lying, would no longer form a part of the business code! At all times, and under all circumstances, to respect the sacredness of life, and the natural rights of man, would become the universal watchword! Justice would dethrone charity! The high moral tone of the industrial and commercial world, would pervade the social and political. The injury of the weakest, would become the concern of the strongest. The rising tide of humanitarianism would submerge poverty. The fires of ignorance and crime, would be extinguished by its conquering flood.

"Than this, no lesson more important, could be taught to the people. The scales of selfishness having fallen from their eyes, they can be made to understand, that all of these wonderful things may be accomplished, quickly and easily, by the plain, practical methods of unselfish co-operation. Methods, whose assured results are as easily demonstrable, as the solution of a mathematical problem. Once convinced, they will make haste to discard the wasteful methods of the competitive system; substituting therefor, the co-operative conservation of national wealth. In this conservation, the wealth of the unit, will be the measure of the wealth of the nation.

"This conservation will usher in a new era, of the means of gathering, and of the higher uses of national wealth. A magnificent national fund, accumulated for the benefit, education, refinement and enjoyment of all. The swiftness of its accumulation and the magnitude of its billions, will become the marvel of the world! By contrast, all former standards of the wealth of nations, will fade and shrink to insignificance! Why must this prove true? Because, under the beneficent reign of co-operative equality, money, shorn of its power, would only be valued for its use. The store of national wealth, being for the equal use and benefit of every individual citizen; the incentive for its accumulation, would inspire all alike. As a result, the people as a mass would enjoy all the benefits of great wealth, minus its burdens, abuses, temptations and dangers. In this, any one of them might be envied by the competitive millionaires.

"Among the many lessons in addition to those enumerated, which the Crusade must teach to the people; I would strongly emphasize the following:

"That human life, as the flower and fruit of the planet—each individual being a microcosm of the macrocosm—must always be held as the most sacred and the most precious of all things. Because it is the object and purpose, the beginning, the expression, the commandment and the fulfillment of the law.

"That the law of life and the law of progress, are complements of each other. Like twin sisters, they act as a bond between the systems of the universe: they embrace all things, from an atom to the Infinite!

"That activity, is the expression of life! Necessity and glory, are the two poles of human activity; its inspiration and its motor power!

"It is the evident purpose of natural law, that the activity of man shall unceasingly produce for all, an abundance of the necessities, comforts and luxuries of life.

"Ignorance, is the giant who bars the pathway of progress! Labor from necessity, reigns as a rule, in all ages of ignorance! Misery and poverty, are its children!

"Labor for glory, marks the age of enlightened progress, where all may have an opportunity to express individuality, through their handiwork; to taste the great joy, that comes with the consciousness of participation in spontaneous, unselfish, intelligent activity, which shall insure the reign of perpetual peace and plenty. In this, man's conquest over matter, becomes the true glory of labor! In the variety of self-chosen, self directed, co-operative, productive labor, is found life's greatest blessing.

"Organized, unselfish co-operation, will teach the people to appreciate the dignity, and the true nobility of labor. From it, they will learn that labor, however simple or insignificant, is far nobler than any kind of enervating idleness; no matter how much that idleness may be gilded by the varnish of honor! Godin says: 'A day's work well done, is worth more than a whole existence of inactivity!'

"Labor develops the possibilities of life! It is the

effective instrument which makes possible the progress of nations, the emancipation of peoples! The labor of passing ages has evolved a fund of ideas, best adapted to guide humanity towards a true interpretation of the object and purpose of human life.

"Labor will cease to be a burden, when man comprehends its true mission. Stripped of its drudgery, released from the harness of toil and the spur of necessity, the brightness of the blessing of labor shines forth resplendent. In the halo of this radiant truth, can anyone be guilty of a blasphemy, which degrades labor to the penalty of a punishment.

"The question of politics is intimately associated with the question of labor. The science of politics, is the science of life. Government, is its expression. Self-government by the individual, is its keynote. The study of this science should be pursued by all classes, with the enthusiasm born of a religious zeal. A few of its most important principles may be found embodied in the following propositions. If we wish to be able to take an interest in moral life; we must first satisfy the demands of physical life. If we wish to practice justice, we must first learn the law of Right and Duty; that is, in striving to satisfy our own material wants, we must learn how to protect the rights of others. We must remember, that they too are toiling for the same purpose.

"In order to protect the welfare of each political unit, these principles must form the basis of all scientific politics. In the social units evolved by cooperative life, these conditions are embodied and expressed. In them, we shall find the basis upon which to build a grand, social, industrial and political organization. An organization, which shall truly represent Liberty and Justice; which, in its expression as a whole, shall be the government of the New Republic!

"Co-operation is the foe of despotism! Associated, intelligent, political co-operation, is the educator which shall teach the people, that a true republic cannot exist until, in the minds of its leaders, every vestige of the spirit of despotism has been cast out.

"In the accomplishment of this great political work, faith in the destiny of this republic, its people, and its mission, is to prove a most important factor. To endow a people with faith, is to multiply their strength tenfold! Faith, reinforced by knowledge, is an irresistible force, against which naught can prevail! Hence, it becomes imperative, that in each school and kindergarten of the republic, its children should be taught in broad outlines, the vastness of its territory, and the magnitude of its natural resources.

"I cannot too strongly emphasize the necessity for this important part of the political education of children! As the future guardians and law makers of the republic, its children should acquire a thorough knowledge of the widely diversified characteristics of each geographical sub-division. This, they must accomplish, before they can be prepared to appreciate the overshadowing significance, of its past, present, and future destiny.

"The kindergarten offers perfect conditions, for the introduction of a primary course of this political instruction. By using a large outline map, showing the geographical and geological formation, the mineral deposits, the extent or area of timbered and agricultural lands, the manufacturing centers, the principal wagon-roads and lines of transportation, the natural trade centers, the population, the schools, the chief officers, and the well known political leaders of each sub-division; a series of intellectual excursions could be so arranged, and made so interesting to the children, that they would soon master these statistics, as identified with every State and Territory in the Republic. Having finished the subdivisions, attention could then be given to a much larger map of the United States, on which the States and Territories on a smaller scale, would show the same statistics. From this map, the study of the political statistics of the States and Territories, by groups, could then be commenced.

"A comparative study of the groups, would be full of interest for the children, and would offer a great number of delightful surprises. The six groups in natural order, should be classified as follows: The New England, the Middle, the Southern States; the States of the great basin of the Mississippi Valley, including the imperial State of Texas; the Rocky Mountain States, and the States of the Pacific Slope, including that remarkable, and only partially explored Territory, Alaska.

"From these group studies, the children may learn many object lessons, which might demonstrate to them,. the natural supremacy of this republic, over other nations. I may mention the following, as noteworthy: The Great Lakes of the Middle West; with a coast line of more than three thousand miles in length; with an interstate commerce which exceeds in tonnage, the combined shipping trade of France and Germany. The marvelous capacity of the great agricultural States of the Mississippi Valley to become the granary of the world; to furnish its entire food supply, of bread, beef and pork. The imperial State of Texas, with its wealth of wheat, cane, corn, cotton and cattle; with a domain so wide, that it equals in extent, that of Great Britain, European Turkey, Switzerland, Denmark and Portugal. Again, passing to the uttermost regions of the Great Northwest, we should find the mammoth Territory of Alaska, rich in its unexplored forests, mineral deposits and golden sands; with a picturesque coast line of fabulous extent, stretching away to the North far beyond the Arctic Circle, indented by a multitude of romantic bays and inlets, where jutting crags, bold promontories of basaltic rock, countless islands, sparkling water and shining glaciers, fill the measure of beauty and grandeur.

"Thus educated, the future guardians of the political welfare of the republic, would understand the natural wants of its widely separated sub-divisions; they would fully appreciate the significance of its destiny as a nation. They would always be loyal to the demands of that destiny, which should be commensurate with its inexhaustable resources, with the magnitude of its domain. A domain so immense, that when compared with the countries of the Old World, without counting island possessions, or the Territory of Alaska, it exceeds in extent, the combined areas of China proper, Japan, Austria, Germany, France, Spain, Portugal, Italy, Holland, Belgium, Greece, Denmark, Norway, Sweden, Switzerland, Great Britain, and European Turkey. With the hearts of its voters inspired by such patriotic teachings, the Republic must endure; must fulfill its prophetic destiny! Naught can prevail against it! Not even the selfish schemes of a corrupt oligarchy; no matter how boldly they plan or how many billions of capital they may control!

"In teaching these things, my children; also in enlarging and perfecting the work of the Crusade, I can promise you the support and co-operation of the spirit world. The broad outlines, which I have given, will suggest the more complete details of the work, which I now leave in your hands."

"That thought alone, Mr. Fenwick," said Fillmore, "ought to prove a tower of strength to us. May we not make that co-operation more effective, by a closer study of the conditions that prevail, and of the laws which govern spirit life?"

"Later on my son, that will be advisable. But just at present, it is of the utmost importance, that every effort should be made to improve the social, industrial, mental and physical condition of mortals, as the necessary foundation for true spiritual growth.

"Mental growth must precede the spiritual. Power exercised by the mind over the body, in moulding physical structure, multiplies the power of the spirit acting on matter, again reacting on both mind and body. Consciousness, is spiritual life. To enlarge the sphere of consciousness, is to add to spiritual growth. Evolution, is nature's effort towards progression. The new spiritual era, which began with the last half of the nineteenth century, was marked by a dawning consciousness in the mind of man, that he might become a self-directing factor in his own evolution. This consciousness in turn, became the starting point of spiritual evolution on the mortal plane of existence. The last, having been made possible by the first.

"Reasoning from the premises stated, we must logically conclude that the embodiment of more mind, of better mind, is a matter of the utmost importance to the whole human race. As body and brain are working parts of the mind, its machinery of expression; it is equally important, that both mind and body should be perfected together. Hence, the necessity for better social conditions, more financial independence, less labor, more leisure, longer life and larger brain capacity; and finally, as the crowning requirement, to be well born! To banish poverty, is to make these things possible.

"Before a proper conception of the spiritual world can be entertained by mortals, their minds, by the aid of the sciences, must have acquired such knowledge of their environments, as shall satisfy the requirements of spiritual evolution. Every item of real knowledge thus gained, is just so much added preparation towards the understanding of the spiritual; towards a harmonious interblending, and cooperation of the two worlds. In accordance with the law of progression, truth, to the ever changing stages of consciousness, is relative. In order to illustrate the relativity of truth, and the magnitude of the domain of knowledge in the mortal state, which must be conquered before consciousness can be extended beyond the confines of the spiritual; let us consider the following, somewhat approximate postulates.

"Let us suppose, that the life of the planet, Earth, embraces all forms of life; each individual life pulsating in harmony with the great mother heart of the planet.

"Let us suppose, that spirits, both embodied and disembodied, incarnate and excarnate, considered as a mass, may act as the terrurgic spiritual body and brain of the planet; subjective and responsive to the inspiration and guidance of the universal cosmic mind, acting from the cosmic center.

"Let us suppose, that the material world, with the atom as its smallest unit, is the medium of mortal existence. Again, that the impalpable ether of the interstellar spaces, is the medium of existence for the spiritual world. And again, as a measure of the fineness of ether, that the difference between an ether particle and an atom, should be as wide as the difference between the atom and the planet.

"Considering these posits as a basis for comparing life in the two realms, we at once perceive that life, organized to correspond with the coarse meshes of the material plane of existence, can be permeated, filled and quickened, by organized spiritual life, without disturbing the unity of either organization. The interblending of spirit and matter, is accomplished. The mystery of the dual existence of soul and body, is explained. The soul in the body, yet, not of the body! The permanent and the enduring, mated with the changing and the ephemeral! The cell life of the physical, with the soul life of the eternal!

"In comparing the two states of existence, the physical with the spiritual, we find the horizon of consciousness in the former, is vaguely defined and very much limited; while in the latter, it is sharply defined and widely extended. The more we study and compare, the more readily we understand, that space, duration, size, minuteness, solidity and porosity, are all relative terms which depend for their significance entirely on the standpoint of consciousness. So apparent is this fact, that we soon learn how impossible it is for the mortal mind to understand, even the more simple elements of spirit life, until the dual or spiritual mind, with its consciousness, has grown and unfolded to the required extent. Hence, growth of consciousness, is growth of spirit; the spirit which molds and controls matter.

"Self-conscious consciousness, is the immortal ego! As a part of the progressive, all inclusive, spiritual life of the planet, it takes part in the evolution and progression of the mass. This mass, in the fulfillment of the purpose of existence, is subjective and responsive to cosmic law, and to cosmic inspiration.

"In these postulates, we have the key which unlocks the mystery of life. We catch a glimpse of its true meaning, purpose, glory and grandeur. They raise the theory and practice of human progress to a question of the first magnitude; to a science of life, which demands the attention of every student. The school of human life, lies at the base of the curiculum of knowledge. It becomes the foundation of spiritual progress, as well. Hence, the importance of rightly cultivating the mind, of extending its consciousness to the uttermost limits of human capacity.

"Selfishness and despotism, are frowning barriers across the pathway of human progress. They thrive by war. War, is the foe of spirituality, the mother of murder! War must be abolished, before man can hope for true spiritual evolution! It is the fortunate destiny of this republic, to lead the race in a crusade against it; to open the way for its final abolition. It is to be the province of the Crusade to teach the people, that war has been the scourge of humanity since the beginning of the historical era; the greatest crime ever perpetrated against the sacredness of human life! Peace, multiplies the products of labor.

Labor, is the genius of life! War, destroys the laborer and his product. War is the genius of death! War, is a symbol of barbarism; it is both the throne and the refuge of despotism. For the purpose of maintaining despotism, people for centuries have been subjected to the hard conditions of unremitting toil, that they might endure the fatigues of war without a murmur. For the same reason, despots have kept the masses in ignorance, lest they should discover the true quality of justice; the moral law, which condemns both despotism and war; lest they should come to realize all the horrors of the most outrageous crime possible to the conception of human reason; the crime of war! War is such an overwhelming calamity, that it is almost impossible to estimate the ruin and the destruction which it has wrought! If the millions of lives and the billions of treasure spent in the world's wars, had been employed in protecting the people, in generating, rearing, sustaining and developing them to the highest attainable point, this earth would now witness a social millennium; where peace and prosperity, high culture and harmonious brotherhood, would reign supreme!

"I rejoice, that I am permitted to prophesy its downfall! Long before the close of the twentieth century, standing armies will disappear; war will be at an end; the angel of peace will spread her white wings over all the nations of the earth! This Crusade, is the beginning of the end! For the encouragement of our Crusaders, I will indicate two causes, acting from opposite directions, which will serve to hasten war's dissolution.

"First: The competitive system, for centuries, has been war's chief recruiting office. Under its reign, in the fierce struggle for existence, it has kept up a perpetual warfare between man and man; always the stronger against the weaker. When vanquished, the weaker as a last resort, could and did, enlist as a soldier. Thanks to the co-operative farm, spread broadcast by the Crusade; the early substitution of the co-operative, for the competitive system, will make the weak strong; make them financially independent! Soldiering as a trade, is made possible by poverty! Whenever a people are emancipated from the cringing slavery of want, naturally averse to being slaughtered, they will rise enmasse, and refuse to be apprenticed to the brutal trade of killing their kind. Thus it will happen, that armies will melt away and disappear, for the want of fighting men!

"Second: Strange as it may appear, the inventors of mighty engines of war, of terrible explosives, of deadly missiles, each in turn, more horribly destructive than the others; are all envoys of peace; that sweet peace, which shall bring rest, renewed energy, and swift progress, to all classes. Through the multiplied and combined efforts of these inventors, the bloody and barbarous art of war, is fast becoming so suicidal, and so financially disastrous to the nations of the earth who have the misfortune to engage in it; that such as wish to preserve a national existence, must do so by making haste to ally themselves with the friends of universal peace, through international arbitration.

"Under such circumstances, the nations of the earth, ground between the inexorable, upper and lower millstones of the first and second cause, acting under pressure of self-preservation, will, with one accord, join in covenanting for a total disarmament, and a perpetual peace. All hail, the glad day!

"Then, will dawn man's era of true spiritual evolution! Then, will the true object and purpose of life, be understood! Then, will the sacredness of human life, be rightly conceived, appreciated, maintained and respected! Then, wholesale murder, no longer sanctioned by man-made laws, it will be possible to banish the spirit of murder from the life of the individual! Then, the lesser crimes, the demons of despotic selfishness, greed, cruelty, and lust for power, which now clog progress and prevent the realization of a practical brotherhood for humanity, can be shaken off and rendered harmless!

"Then, the emancipated legions of toilers, will rise to a true understanding of the blessing of labor as the real expression of life; that the glory of labor, is man's conquest over matter; that food, shelter, raiment, and sustenance for body, mind and soul, are the essential elements of life; a natural equipment for the conquest! Then, it will be the province of a natural religion to teach the people how to help themselves! how to master the great problem of physical life, by attaining the greatest perfection in feeding, clothing, housing, educating, and spiritualizing humanity!

"Then, the solidarity of the spiritual welfare of mankind, will equal that of the physical! Then, the measure of spiritual progress achieved by the mass, will be the measure of progress attained by its weakest unit! Then, will come perfect co-operation, between the spiritual and the physical! Then, will come the reign of liberty and justice, the guardian spirits of a true republic! Then, will come the social, the industrial, and the spiritual millennium! Then, the barriers of selfishness will have been burned away; the two worlds will be united; in the new atmosphere of brotherly love, spirit and mortal may harmoniously walk, talk, and work together for the perfection of the race!

"Then, the great armies of the world, no longer in the guise of organized barbarism, or a tax on the industries of the nations, will be converted into armies of peace, engaged in the production of real wealth! Then, the heretofore undreamed of store of public wealth, will, in its proper distribution, give to all mankind, the acme of universal education, civilization and happiness!"

CONCLUSION.

Born leaders of a progressive age; filled with the inspiration of one great purpose in life; at all times, equal to the demands of the hour; hand in hand, with hearts united by the bonds of a supreme love; nobly unselfish, and spiritually refined; generous, handsome, accomplished; wealthy, eloquent and magnetic; Fillmore and Fern, our hero and heroine, were everywhere recognized as a commanding force in the social and political world. A force which quickly overcame all opposing obstacles. They were so much interested, and so absorbed in the ever increasing success of the Crusade, that the happy months and years flew swiftly by. Their devotion to each other, was a potent charm which begat in the hearts of a legion of admiring followers, an intense loyalty to them, and to the banner of the Crusade, which had led them to so many victories in the cause of humanity.

The second decade of the century was throbbing with the birth of epoch-making events. The astrological forces seemed in conjunction with planetary evolution. The time was ripe for the incoming wave of a new social era. The spirit of progress was brooding in the air; stirring in the hearts of the people, who hailed the Crusaders as blessed evangels of the new life, for which they had yearned and prayed so many years. The gospel of the new life, was the gospel of co-operative labor. The wonderful strength and effectiveness of the co-operative farm movement, to lift the laborer from conditions of ignorance and poverty, to those of financial independence, comfort and refinement; was practically demonstrated, a thousand times over. To the people, each demonstration was an ever growing source of astonishment and delight. The enthusiasm aroused, burning with the fires of a religious zeal, irresistibly drew them into the ranks of this powerful organization. With rapidly increasing numbers, it swept over the land with the force and fury of a great tidal wave! In its track, on the ruins of the competitive system, there was established, the reign of co-operative peace and plenty, the social and political millennium.

Among the leaders of the Crusade, assembled at Washington, George and Gertrude Gerrish were especially prominent. To them was assigned the task of organizing the lecturing or missionary bureau of the Crusade; its trained force of traveling educators. The good work accomplished by this force, was another well earned tribute to their extraordinary skill as organizers. As well fitted for the responsible duties; George Gaylord and Honora Eloise Houghton, having become inseparable friends, engaged lovers, and finally a well-mated, conjugal couple; were placed in charge of the traveling educators on the Pacific Slope. So eloquently and effectively did they labor in this wide field, that throughout its length and breadth, they became very popular, winning hosts of friends for themselves and the cause.

Solaris Farm and village, the working center of the movement, soon doubled many times, its territory and population. It became an important manufacturing center, which made an ideal home for the National Co-operative Farm School; a normal school, which every year graduated teachers by the score. The history of Solaris as the initial farm made it so famous, that thousands of enthusiastic co-operators annually visit it. It is the business of the reception committee appointed by the normal school, to receive, entertain and instruct these visitors.

Gilbert Gerrish, true to his arisen sweetheart, and to his own peculiar purpose in life; declined to leave Solaris, with his parents. Indeed, he was so universally beloved by its young people, that they could not, and would not give him up! To the visiting stranger, he seems by far the most popular and the most highly honored young man in the village. This distinguished consideration, he has rightfully and honestly earned. Happy himself, in generously using his rare gifts for making other people happy!

Thus endeth the story of Solaris Farm. May its purposes haunt the minds of its readers, like the memories of some prophetic dream, which may not be obliterated, which can not be forgotten.

A FEW POINTERS FROM THE PEN OF THE REVIEWER.

Solaris Farm is the title of a new book "with a purpose." In fact it is a book with many purposes. While the author writes intelligently and forcefully upon stirpiculture, education, invention, hygiene, sanitation, moral, physical and mental growth and culture, and injects many new, beautiful and practical thoughts into each of these subjects, his chief theme is unselfish co-operation, his chief purpose is to exhibit the benefits, moral, phyiscal, social and financial, that will be showered upon the human family when they become wise enough to cease competing with each other, and progressive enough to begin co-operating.

The story is the logical development of the following situation: Fern Fenwick, an heiress to a vast estate, had promised her father before his death to use a good share of the Fenwick millions in bettering the condition of the race. Her first experiment is a co-operative farm of about five thousand acres, whereon about two hundred and fifty families settle and work out the many problems which the author desires to discuss.

In all of these operations she has the able assistance of Fillmore Flagg, a farmer's son, who, having seen his father and dozens of his old neighbors crushed in spirit and broken in fortune by the resistless trend of events under the competitive system with all its waste of misdirected energy, has become disgusted with the meager results of farm work and having by great energy obtained a practical education has determined to do something for the alleviation of the miseries of a competition crushed society.

He meets Fern Fenwick and is by her employed to superintend the co-operative farm.

A very pretty little love story, which the author has told with pleasant humor, is the result of their meeting, but the weightier themes with which the book is filled are likely to more fully engross the attention of the reader.

Co-operative ventures have usually been founded upon some "ism," and were held together by its religious or other influence. In the Solaris Farm colony a very comprehensive scheme of insurance against accident, poverty, sickness and old age is the binding principle. The premium is the profit which the co-operators collectively make by producing what they want (or by buying at wholesale what they cannot produce) and selling the same to themselves individually at regular market rates. The excellence of their wares attract many purchasers from the outside and the profits resulting therefrom also tend to swell the insurance fund of the co-operators.

All kinds of business, and manufacturing are carried on by the co-operators in addition to farming. Co-operative thinking solves the knottiest problems for the colony, invention flourishes and, once started, money flows into their coffer at a fairly satisfactory rate.

Co-operation is the key-word, the essence, the very soul of Solaris Farm. All the successes achieved by the characters that people the book are the results of co-operative working, thinking and saving. Every stockholder lends a hand, and lo! the hours of labor are short and delightful; when a disagreeable task must be done, co-operative thinking invents a machine which does the work better than a man could do it; the dignity of toil is established on a sure foundation, and the statement that "muscular effort is a mental demonstration," is verified.

"Will it pay?" is sometimes called "the American question." In Solaris Farm the author has successfully undertaken to present an unselfishness that will pay—not in the fairy gold of a far-off Heaven, but in the coin of the realm, here and now. Leisure for study and recreation; books, pictures, objects of beauty and art; better health; longer life; the society of delightful people none of whom are competing for the lion's share, but all of whom are co-operating for the benefit of the community; absence of the fear of poverty; certainty of support in sickness and old age;—all these and thousands of other comforts are some of the certain wages of unselfishness.

A feature of Solaris Farm which will commend itself to every well-wisher of the race is the high estimate which the author places on humanity. Man, he says, is the flower and fruit of the planet, its highest and best product. To arrive at the highest point possible in his evolution, it is necessary for him to be well born and this necessitates happy, healthy, prosperous parents and proper environments. To follow out this idea to its logical conclusion would be to repeat the author's arguments, for he has completely filled the field. The reader is referred to the story for the facts proving that unselfish co-operation will furnish everything needful for the complete unfoldment of the now almost dormant possibilities of human nature.

The pursuit of happiness and the hope of its ultimate possession is the motor which induces all human endeavor. No act is ever done except in obedience to this law of our nature which compels us to seek pleasure. Ignorance of the nature of true pleasure has led us after many a will-o'-the-wisp, and our unlearned race has soiled its garments many times in error, commonly called "sin." "Sinful pleasures," against which our parents, the clergy, and all moral philosophers have warned us, do not exist. *There is no pleasure in sin.* Our race beliefs, based upon untruth and ignorance, have bequeathed us a heritage of appetites, passions and desires which are wrong, and hurtful when gratified.

Among the most hurtful of race beliefs is the fixed idea that labor is a curse. Nothing could be further from the truth. As has been aptly said: "Art is the expression of a man's joy in his work." Labor—muscular exertion, having a definite productive object—is a blessing and a joy when the worker is in love with his work. Work is a curse only under the competitive system, which by its wasteful methods extends the hours of toil beyond the limits of endurance, robs the worker of the full benefits of his labor and gives him no time for self-improvement. The experience of the stockholders of Solaris Farm shows how the ancient curse was removed by unselfish co-operation, and labor crowned with the dignity that is its due.

While Solaris Farm was not intended as a propaganda of spiritualism, that cult has been introduced with considerable dramatic effect for two apparent reasons. The first and least important of these reasons is to cater to the ever-growing taste of the reading public for the occult; but the second reason is peculiar to the book. In discussing man as the most valuable product of the planet, and the relation which the soul bears to the body, it became necessary to approach the subject from the view-point of one who is in nowise affected by the petty altercations, jealousies and strifes of the world; one who knows by experience all the hardships of life and its many temptations, but who has also progressed beyond the sphere of their influence. The most natural and obvious way of obtaining this coveted point of observation was to let the spirit of such a noble character as Fennimore Fenwick speak from the fullness of his experience, both as mortal and spirit, of the needs of the race, the curse of competition, the value of proper environmental conditions for perfect motherhood, pre-natal education and adequate training of mind and body, such as may not be secured even by the most wealthy in the present condition of society, but which would be the heritage of every individual in a co-operative community. The utterances of Fennimore Fenwick rank with the best thought on these subjects and no person can read them without having implanted in his breast a higher regard for his race, and a greater solicitude for the material and spiritual unfoldment of humanity.

For many years, orators and agitators have vied with each other in proclaiming that capital and labor were the two factors of financial success. They were and still are mistaken. Within the pages of Solaris Farm the reader is given the true formula, which may be algebraically stated thus: "Capital + Labor + Brains = Financial Success." Financial Success, however is not the complete product of these factors when selfishness, greed and wasteful competition are eliminated from the equation by the substitution of unselfish co-operation. The happy result of the experiment at Solaris Farm must convince the reader of the correctness of the formula and the value of the substitution.

In considering the broad field covered by this attractive book; its wide departure from the mission of the ordinary novel, its probable use as a text-book of advanced thought on true socialism, progressive co-operation, a new order of political economy and the ways and means of making colony life desirable, successfully coherent, self-supporting and practically delightful; the price of Solaris Farm (50 cts, in paper covers, $1.25 in cloth binding) will commend itself to the purchaser as not only reasonably moderate, but also if he be an interested reader, with business intentions, that the large end of the bargain is very much in his favor.

Solaris Farm was written by Captain Milan C. Edson, whose military title was earned during the great Civil War. He was a farmer and the son of a farmer. He enlisted as a private soldier and without influence rose to a captaincy by merit and bravery alone. He is a profound thinker, a lover of his race and has given many years to the study of social and political questions. It has been his desire to found a community where his ideas of true success might be wrought out, as an object lesson to the world, of the advantages of unselfishness. This pleasure having been denied him, he has incorporated his leading ideas in Solaris Farm, in the hope that some one more fortunate than himself may be able to receive the blessings which must inevitably flow from such a noble life.

Utopian Literature

AN ARNO PRESS/NEW YORK TIMES COLLECTION

Adams, Frederick Upham.
President John Smith; The Story of a Peaceful Revolution. 1897.

Bird, Arthur.
Looking Forward: A Dream of the United States of the Americas in 1999. 1899.

[Blanchard, Calvin.]
The Art of Real Pleasure. 1864.

Brinsmade, Herman Hine.
Utopia Achieved: A Novel of the Future. 1912.

Caryl, Charles W.
New Era. 1897.

Chavannes, Albert.
The Future Commonwealth. 1892.

Child, William Stanley.
The Legal Revolution of 1902. 1898.

Collens, T. Wharton.
Eden of Labor; or, The Christian Utopia. 1876.

Cowan, James.
Daybreak. A Romance of an Old World. 1896. 2nd ed.

Craig, Alexander.
Ionia; Land of Wise Men and Fair Women. 1898.

Daniel, Charles S.
AI: A Social Vision. 1892.

Devinne, Paul.
The Day of Prosperity: A Vision of the Century to Come. 1902.

Edson, Milan C.
Solaris Farm. 1900.

Fuller, Alvarado M.
A. D. 2000. 1890.

Geissler, Ludwig A.
Looking Beyond. 1891.

Hale, Edward Everett.
How They Lived in Hampton. 1888.

Hale, Edward Everett.
Sybaris and Other Homes. 1869.

Harris, W. S.
Life in a Thousand Worlds. 1905.

Henry, W. O.
Equitania. 1914.

Hicks, Granville, with Richard M. Bennett.
The First to Awaken. 1940.

Lewis, Arthur O., editor
American Utopias: Selected Short Fiction. 1790–1954.

McGrady, Thomas.
Beyond the Black Ocean. 1901.

Mendes H. Pereira.
Looking Ahead. 1899.

Michaelis, Richard.
Looking Further Forward. An Answer to *Looking Backward* by Edward Bellamy. 1890.

Moore, David A.
The Age of Progress. 1856.

Noto, Cosimo.
The Ideal City. 1903.

Olerich, Henry.
A Cityless and Countryless World. 1893.

Parry, David M.
The Scarlet Empire. 1906.

Peck, Bradford.
The World a Department Store. 1900.

Reitmeister, Louis Aaron.
If Tomorrow Comes. 1934.

Roberts, J. W.
Looking Within. 1893.

Rosewater, Frank.
'96; A Romance of Utopia. 1894.

Satterlee, W. W.
Looking Backward and What I Saw. 2nd ed. 1890.

Schindler, Solomon.
Young West; A Sequel to Edward Bellamy's Celebrated Novel "Looking Backward." 1894.

Smith, Titus K.
Altruria. 1895.

Steere, C. A.
When Things Were Doing. 1908.

Taylor, William Alexander.
Intermere. 1901.

Thiusen, Ismar.
The Diothas, or, A Far Look Ahead. 1883.

Vinton, Arthur Dudley.
Looking Further Backward. 1890.

Wooldridge, C. W.
Perfecting the Earth. 1902.

Wright, Austin Tappan.
Islandia. 1942.